# UNITEXT for Physics

UNITEXT for Physics series, formerly UNITEXT Collana di Fisica e Astronomia, publishes textbooks and monographs in Physics and Astronomy, mainly in English language, characterized of a didactic style and comprehensiveness. The books published in UNITEXT for Physics series are addressed to graduate and advanced graduate students, but also to scientists and researchers as important resources for their education, knowledge and teaching.

More information about this series at http://www.springer.com/series/13351

Michele Livan · Richard Wigmans

# Calorimetry for Collider Physics, an Introduction

 Springer

Michele Livan
Dipartimento di Fisica
Università di Pavia
Pavia, Italy

Richard Wigmans
Physics and Astronomy Department
Texas Tech University
Lubbock, TX, USA

ISSN 2198-7882 ISSN 2198-7890 (electronic)
UNITEXT for Physics
ISBN 978-3-030-23655-7 ISBN 978-3-030-23653-3 (eBook)
https://doi.org/10.1007/978-3-030-23653-3

This Springer imprint is published by the registered company Springer Nature Switzerland AG
The registered company address is: Gewerbestrasse 11, 6330 Cham, Switzerland

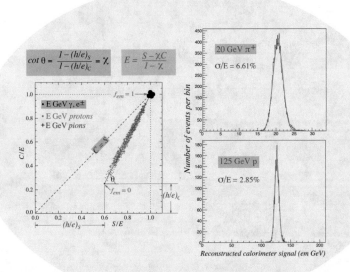

*"Dual-readout calorimetry"*

*For Nadia and Nazzi
with love*

# Preface

In the past 40 years, calorimetry has become an extremely important technique for the study of the subatomic structure of matter. Especially experiments that make use of the increasingly powerful accelerators and storage rings rely heavily on calorimetry. Calorimeters fulfill a number of crucial tasks, ranging from event selection and triggering to precision measurements of the four-vectors of individual particles and jets and of the energy flow in the events (missing energy, *etc.*). This development has benefited in no small part from the improved understanding of the working of these, in many respects somewhat mysterious, instruments.

Much has been learned about calorimetry, primarily thanks to dedicated R&D projects. This information is contained in a very large number of papers, scattered in the scientific literature. Over the years, many review articles have been published, in which the state-of-the-art is summarized. However, these review papers usually concentrate on specific aspects and their educational value is typically rather limited.

We decided to write this book as a compact educational introduction for those students (and others) who are confronted for the first time with calorimetric particle detection. It is based on our experience teaching courses on instrumentation at our universities and at summer schools. We assume a basic understanding of physics, including quantum mechanics and (sub)atomic physics as taught in a typical undergraduate curriculum.

The book is subdivided into five parts. After a general introductory chapter, Part I concentrates on the physics that is relevant for the interactions of particles with matter, for shower development and the signals produced by calorimeters. Part II deals with the aspects of shower development that affect the calorimeter performance in practice. Part III is exclusively dedicated to the specific problems encountered in hadron calorimetry, and the methods that have been developed to mitigate the effects of these problems. Part IV deals with the challenges encountered when operating a calorimeter system in an experiment. This part concerns mainly calibration issues, which are in practice frequently misunderstood and mishandled, even by experienced experimentalists. In Part V, we describe the state of the art in calorimeter performance. We also dedicate in this part a chapter to Particle Flow

Analysis, a concept that has been proposed as an alternative for calorimetric measurements of jets in some experiments.

We wish to thank Barbara Amorese and Marina Forlizzi of Springer who guided us in bringing this project from concept to publication. Finally, it is to the two people who are making our lives so extremely wonderful that we dedicate this book: our wives, Nadia Ranzani and Nazzi Wigmans. Their love, patience, encouragement and support have been essential for its completion.

Pavia, Italy                                                    Michele Livan
Lubbock, USA                                               Richard Wigmans
April 2019

# Contents

1   Calorimetry—From Thermodynamics to Particle Detection . . . . . .   1
    1.1   What Is a Calorimeter? . . . . . . . . . . . . . . . . . . . . . . . . .   1
    1.2   Short History . . . . . . . . . . . . . . . . . . . . . . . . . . . . . . .   2
    1.3   Types of Calorimeters Used as Particle Detectors . . . . . . . . . .   10
    1.4   Techniques Used for Signal Generation in Calorimetry . . . . . . .   14
          1.4.1   Scintillation . . . . . . . . . . . . . . . . . . . . . . . . . .   14
          1.4.2   Čerenkov Radiation . . . . . . . . . . . . . . . . . . . . . .   16
          1.4.3   Ionization . . . . . . . . . . . . . . . . . . . . . . . . . . . .   18
          1.4.4   Cryogenic Phenomena . . . . . . . . . . . . . . . . . . . .   21
          1.4.5   Acoustic Signals . . . . . . . . . . . . . . . . . . . . . . . .   23
    References . . . . . . . . . . . . . . . . . . . . . . . . . . . . . . . . . . . .   26

Part I   The Basics of Calorimetry

2   Interactions of Particles with Matter . . . . . . . . . . . . . . . . . . . .   31
    2.1   Introduction . . . . . . . . . . . . . . . . . . . . . . . . . . . . . . . .   31
    2.2   The Electromagnetic Interaction . . . . . . . . . . . . . . . . . . . . .   31
          2.2.1   Electrons and Positrons . . . . . . . . . . . . . . . . . . . .   32
          2.2.2   Muons Traversing Dense Material . . . . . . . . . . . . . .   35
          2.2.3   Photon Interactions . . . . . . . . . . . . . . . . . . . . . .   37
    2.3   The Strong Interaction . . . . . . . . . . . . . . . . . . . . . . . . . .   41
          2.3.1   Particle Production in the Absorption
                  of High-Energy Hadrons . . . . . . . . . . . . . . . . . . . .   42
          2.3.2   Nuclear Reactions in the Absorption
                  of High-Energy Hadrons . . . . . . . . . . . . . . . . . . . .   43
          2.3.3   The Interactions of Neutrons with Matter . . . . . . . . . .   49
    References . . . . . . . . . . . . . . . . . . . . . . . . . . . . . . . . . . . .   51

**3    Shower Development** . . . . . . . . . . . . . . . . . . . . . . . . . . . . . . . . .    53
    3.1     Electromagnetic Showers . . . . . . . . . . . . . . . . . . . . . . . . . .    53
        3.1.1     Differences Between Electron and $\gamma$ Induced
                Showers . . . . . . . . . . . . . . . . . . . . . . . . . . . . . . .    54
    3.2     Hadron Showers . . . . . . . . . . . . . . . . . . . . . . . . . . . . . . . .    56
    3.3     Material Dependence . . . . . . . . . . . . . . . . . . . . . . . . . . . . .    59
        3.3.1     The Radiation Length . . . . . . . . . . . . . . . . . . . . . .    59
        3.3.2     The Molière Radius . . . . . . . . . . . . . . . . . . . . . . .    60
        3.3.3     The Nuclear Interaction Length . . . . . . . . . . . . . . . .    62
        3.3.4     The Ratio $\lambda_{\mathrm{int}}/X_0$ . . . . . . . . . . . . . . . . . . . . . . . .    63
    3.4     The Importance of the Last Stages of the Absorption
        Process . . . . . . . . . . . . . . . . . . . . . . . . . . . . . . . . . . . . . .    64
    3.5     The Sampling Fraction . . . . . . . . . . . . . . . . . . . . . . . . . . .    67
        3.5.1     Minimum Ionizing Particles . . . . . . . . . . . . . . . . .    67
        3.5.2     Sampling of Non-mips . . . . . . . . . . . . . . . . . . . . .    68
    References . . . . . . . . . . . . . . . . . . . . . . . . . . . . . . . . . . . . . . . . .    73

**4    The Calorimeter Signals** . . . . . . . . . . . . . . . . . . . . . . . . . . . . . .    75
    4.1     Introduction . . . . . . . . . . . . . . . . . . . . . . . . . . . . . . . . . . .    75
    4.2     Signal Linearity and Non-linearity . . . . . . . . . . . . . . . . . . .    75
        4.2.1     Non-linearity Resulting from Quenching Effects . . . . . .    78
        4.2.2     Non-linearity Resulting from Saturation . . . . . . . . . . .    80
    4.3     The Texas Tower Effect . . . . . . . . . . . . . . . . . . . . . . . . . . .    83
    4.4     Signals from External Sources . . . . . . . . . . . . . . . . . . . . . .    87
    References . . . . . . . . . . . . . . . . . . . . . . . . . . . . . . . . . . . . . . . . .    89

**Part II    Shower Aspects Important for Calorimetry**

**5    Containment and Profiles** . . . . . . . . . . . . . . . . . . . . . . . . . . . . . .    93
    5.1     Introduction . . . . . . . . . . . . . . . . . . . . . . . . . . . . . . . . . . .    93
    5.2     Electromagnetic Showers . . . . . . . . . . . . . . . . . . . . . . . . . .    94
        5.2.1     Shower Containment . . . . . . . . . . . . . . . . . . . . . . .    94
        5.2.2     Shower Profiles . . . . . . . . . . . . . . . . . . . . . . . . . .    95
        5.2.3     Experimental Data . . . . . . . . . . . . . . . . . . . . . . . .    98
        5.2.4     Dependence on the Detection Mechanism . . . . . . . . . .    99
    5.3     Hadronic Showers . . . . . . . . . . . . . . . . . . . . . . . . . . . . . . .    101
        5.3.1     Longitudinal Shower Containment . . . . . . . . . . . . . . .    101
        5.3.2     Shower Profiles . . . . . . . . . . . . . . . . . . . . . . . . . .    102
        5.3.3     Lateral Shower Containment . . . . . . . . . . . . . . . . . .    104
        5.3.4     Dependence on Hadron Type . . . . . . . . . . . . . . . . . .    105
        5.3.5     Dependence on the Detection Mechanism . . . . . . . . . .    107
    5.4     Application of Differences . . . . . . . . . . . . . . . . . . . . . . . . . .    109
    References . . . . . . . . . . . . . . . . . . . . . . . . . . . . . . . . . . . . . . . . .    110

**6    The Energy Resolution of Calorimeters** ..................... 111
6.1     Introduction ........................................ 111
6.2     The Effects of Fluctuations on the Calorimeter
        Performance ....................................... 112
6.3     Signal Quantum Fluctuations........................ 114
        6.3.1     Semiconductor Crystals ................... 115
        6.3.2     Cryogenic Detectors ...................... 115
        6.3.3     Čerenkov Calorimeters .................... 116
        6.3.4     Calorimeters Based on Scintillation ....... 119
6.4     Sampling Fluctuations.............................. 120
6.5     Measuring the Contribution of Different Types
        of Fluctuations ................................... 123
        6.5.1     SPAKEBAB ............................... 123
        6.5.2     ZEUS ................................... 124
6.6     Shower Leakage ................................... 128
        6.6.1     Effects of Leakage on the Calorimetric Quality ...... 129
        6.6.2     The Different Types of Leakage ............ 129
6.7     Instrumental Effects ............................... 133
        6.7.1     Electronic Noise.......................... 133
        6.7.2     Variations in Sampling Fraction ........... 134
        6.7.3     Non-uniformity of Active Elements ........ 135
6.8     Misconceptions Affecting the Measured Energy Resolution .... 137
References ............................................... 140

**Part III   Hadron Calorimetry: The Problems and How to Solve
            Them**

**7    The Fundamental Problems of Hadron Calorimetry** ............ 143
7.1     Introduction ........................................ 143
7.2     The $e/h$ Ratio and Its Consequences................... 143
7.3     The $e/mip$ Ratio and Its Effects for Jet Detection ........... 151
7.4     The $e/mip$ and $e/h$ Ratios for Different Types
        of Calorimeters ................................... 152
7.5     Longitudinal Segmentation of Non-compensating
        Calorimeters...................................... 154
7.6     Summary ......................................... 154
References ............................................... 155

**8    Methods to Improve Hadronic Calorimeter Performance** ........ 157
8.1     Introduction ........................................ 157
8.2     Compensation..................................... 157
8.3     Dual-Readout Calorimetry .......................... 161
        8.3.1     Dual-Readout Analysis Procedures ......... 162
        8.3.2     Some Experimental Results................ 164

8.4     Dual-Readout Versus Compensation . . . . . . . . . . . . . . . . . . . . .   166
8.5     Exploiting the Time Structure of the Signals . . . . . . . . . . . . . .   173
References . . . . . . . . . . . . . . . . . . . . . . . . . . . . . . . . . . . . . . . . . . . .   175

**Part IV    Challenges Encountered in Practice**

**9    Calibrating a Calorimeter System** . . . . . . . . . . . . . . . . . . . . . . . .   179
9.1     Introduction . . . . . . . . . . . . . . . . . . . . . . . . . . . . . . . . . . . . . .   179
9.2     Longitudinally Unsegmented Systems . . . . . . . . . . . . . . . . . . .   180
9.3     Longitudinally Segmented Systems . . . . . . . . . . . . . . . . . . . . .   180
        9.3.1    The Basic Problem . . . . . . . . . . . . . . . . . . . . . . . . . .   181
        9.3.2    The HELIOS Calorimeter . . . . . . . . . . . . . . . . . . . . .   183
        9.3.3    Intercalibration with em Showers . . . . . . . . . . . . . . .   186
        9.3.4    Three Compartments—The ATLAS LAr
                 Calorimeter . . . . . . . . . . . . . . . . . . . . . . . . . . . . . . .   188
        9.3.5    Many Compartments—The AMS Calorimeter . . . . . . .   190
        9.3.6    Intercalibration with Hadronic Showers . . . . . . . . . . .   192
        9.3.7    Each Section Calibrated with Its Own Particles . . . . . .   193
        9.3.8    Forcing Signal Linearity for Hadron Detection . . . . . . .   195
        9.3.9    No Starting Point Dependence of Hadronic
                 Response . . . . . . . . . . . . . . . . . . . . . . . . . . . . . . . .   196
        9.3.10   Dummy Compensation . . . . . . . . . . . . . . . . . . . . . . .   198
        9.3.11   The Right Way . . . . . . . . . . . . . . . . . . . . . . . . . . . .   198
        9.3.12   Validation . . . . . . . . . . . . . . . . . . . . . . . . . . . . . . .   200
9.4     Consequences of Miscalibration . . . . . . . . . . . . . . . . . . . . . . .   201
9.5     Off-Line Compensation . . . . . . . . . . . . . . . . . . . . . . . . . . . . .   205
9.6     Conclusions . . . . . . . . . . . . . . . . . . . . . . . . . . . . . . . . . . . . .   207
References . . . . . . . . . . . . . . . . . . . . . . . . . . . . . . . . . . . . . . . . . . . .   209

**10    Operational Challenges** . . . . . . . . . . . . . . . . . . . . . . . . . . . . . . .   211
10.1    Operation in a Magnetic Field . . . . . . . . . . . . . . . . . . . . . . . .   211
        10.1.1   Construction Materials . . . . . . . . . . . . . . . . . . . . . .   211
        10.1.2   Signal Readout in a Magnetic Field . . . . . . . . . . . . . .   212
        10.1.3   Effects on the Calorimeter Signals . . . . . . . . . . . . . . .   213
10.2    Radiation Damage . . . . . . . . . . . . . . . . . . . . . . . . . . . . . . . .   215
10.3    Pileup . . . . . . . . . . . . . . . . . . . . . . . . . . . . . . . . . . . . . . . . .   217
References . . . . . . . . . . . . . . . . . . . . . . . . . . . . . . . . . . . . . . . . . . . .   219

**Part V    The State of the Art**

**11    Calorimeter Performance** . . . . . . . . . . . . . . . . . . . . . . . . . . . . .   223
11.1    Energy Measurement . . . . . . . . . . . . . . . . . . . . . . . . . . . . . . .   223
11.2    The Other Components of the Four-Vector . . . . . . . . . . . . . . . .   226

11.2.1 The Center-of-Gravity of the Showers ............. 226
11.2.2 Localization Through Timing ................... 230
11.3 Particle Identification .............................. 232
11.3.1 Electron Identification in Practice ............... 232
11.3.2 Longitudinally Unsegmented Calorimeters ......... 235
11.4 Tricks to Obtain Useful Details of the Shower
Development ..................................... 237
11.4.1 Exploiting the Wonderful Features of Čerenkov
Light ...................................... 237
11.4.2 A Caveat .................................. 241
References ........................................... 243

12 Particle Flow Analysis ................................... 245
12.1 Introduction ...................................... 245
12.2 The Importance of Calorimetry for PFA ................ 246
12.3 PFA at LEP, the Tevatron and the LHC ................ 247
12.4 PFA Calorimeter R&D ............................. 249
References ........................................... 256

13 Outlook ............................................... 257
13.1 Calorimeters and Physics Discoveries ................. 257
13.2 Lepton Versus Hadron Colliders ..................... 259
13.3 The Future of Calorimetry .......................... 262
References ........................................... 263

Correction to: Calorimetry for Collider Physics, an Introduction ..... C1

Appendix: Some Data Relevant to Calorimetry ................... 265

# Chapter 1
# Calorimetry—From Thermodynamics to Particle Detection

## 1.1 What Is a Calorimeter?

The term *calorimetry* means "measurement of heat". It was first used in thermodynamics, the branch of physics that relates energy, heat, temperature and work. The *calorie*, a unit of energy, was introduced by Nicolas Clément in 1824. It is defined as the energy needed to increase the temperature of one gram of water by one degree Celsius at a pressure of one atmosphere. Even though the unit is based on the metric system, it has become somewhat obsolete after the introduction of the SI system of units, where it has been replaced by the Joule: 1 cal $=$ 4.184 J.

Figure 1.1 shows examples of calorimeters that can be used for measurements in this context. The isolated container in Fig. 1.1a can, for example, be used to measure the latent heat of ice. When adding 10 g of 0° ice to 100 g of 20° water, the temperature of the mixture is observed to drop to 11 °C. From this, one can conclude that the latent heat of ice ($E_L$) is

$$E_L = (100 \times 9 - 10 \times 11)/10 = 79 \text{ cal/g}$$

Next, the container may be placed in the Sun, and after one hour the temperature has increased from 11° to 38°. The container has thus received $27 \times 110 = 2970$ calories, i.e., 0.825 cal/s or 3.45 J/s. Since the exposed surface area is 30 cm$^2$, this corresponds to 1.15 kW/m$^2$. Of course, such a measurement of the solar constant requires the container to be perfectly isolated, so that all the heat received from the Sun is indeed used to increase the temperature of the water, and for nothing else.

In practice, instruments of this type are therefore a bit more sophisticated (Fig. 1.1b). They are used, for example, to measure the energy released in certain chemical reactions, or the specific activity of radioactive samples. In the latter application, it offers the possibility for a non-destructive assay of fissile materials, such as plutonium samples. For example, the specific power of the isotopes $^{238}$Pu and $^{239}$Pu amounts to 567 and 1.9 mW/g, respectively. A calorimetric measurement may

© Springer Nature Switzerland AG 2019
M. Livan and R. Wigmans, *Calorimetry for Collider Physics, an Introduction*,
UNITEXT for Physics, https://doi.org/10.1007/978-3-030-23653-3_1

**(a)**                                                          **(b)**

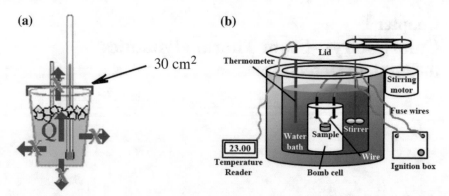

**Fig. 1.1** Examples of thermodynamic experiments in which calorimetry is used. Measurement of the latent heat of melting ice (**a**), and a setup used to measure the energy released in chemical reactions (**b**)

therefore provide information about the relative content of the short-lived (88 yr) $^{238}$Pu isotope and hence on the possible production mechanism of the sample.

In nuclear and particle physics, and more recently in astrophysics experiments, calorimeters are used to measure the energy of particles produced in the reactions that are being studied. There are similarities and differences with the instruments mentioned above. They are similar in the sense that a good measurement requires that the entire energy of the particle is indeed deposited in the sensitive volume of the instrument, any leakage implies a mis-measurement of that energy. The main difference derives from the fact that the energies one wants to measure in these experiments are many orders of magnitude smaller. The most energetic manmade particles are nowadays produced at the Large Hadron Collider at CERN (Genève, Switzerland). The protons accelerated in this machine travel at a speed of 99.999999% of the speed of light and carry a kinetic energy that is about 9000 times larger than the energy contained in their mass ($E = mc^2$). Yet, this kinetic energy is equivalent to only 0.0000003 cal. Even a simultaneous dump of a million such particles in a water container will thus have a negligible effect on the temperature of that water. For this reason, these calorimeters are based on other mechanisms than a temperature increase of the instrument. These mechanisms may include the production of ionization charge, light or sound.

## 1.2 Short History

The first application of calorimetric techniques to detect individual particles and measure their energy took place in the 1940s. It turned out that certain scintillating crystals, such as NaI(Tl), could be used for detecting $\gamma$ rays produced in the decay of radioactive nuclides, and that the energy of such $\gamma$ rays could be measured with

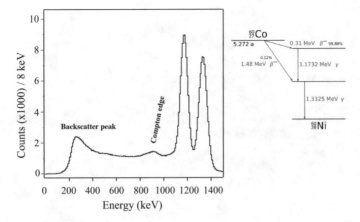

**Fig. 1.2**  Spectrum of $\gamma$ rays from the decay of $^{60}$Co, measured with a scintillating CsI(Na) crystal

a reasonable precision in this way. Before that, A crucial aspect of the new detector technique was the use of photomultiplier tubes (PMTs), which made it possible to convert the scintillation light produced by the crystals into electric signals.

Figure 1.2 shows an example of a $\gamma$ ray spectrum from the decay of $^{60}$Co, measured with a CsI crystal doped with sodium. The two peaks represent events in which the total energy carried by the 1.17 and 1.33 MeV $\gamma$s was deposited inside the crystal. The background "ridge" separating the "backscatter peak" and the "Compton edge" is the result of events in which some energy leakage out of the crystal occurred. Details about the various processes that play a role in this are given in the next chapter.

One of the interesting aspects of Fig. 1.2 is the fact that the two $\gamma$s are clearly recognized as such. In spite of the fact that their energies differ by only 13%, the two peaks are almost completely resolved by the detector, which is thus said to have a *good energy resolution*. As we will see later, there are $\gamma$ ray detectors with an even much better energy resolution, but this feature of the scintillating crystals generated the interest of scientists working on experiments in which particles with much higher energies than carried by these nuclear $\gamma$s were produced.

An example of an experiment where crystals were applied was E-70 [1], one of the first experiments carried out with the 200 GeV proton accelerator (a world record at that time) at Fermilab in the early 1970s. Like many experiments in those days, the experimental setup consisted of a magnetic spectrometer with two arms (Fig. 1.3). These arms consisted each of a series of wire chambers, which were used to track the charged particles produced in the target. The momentum of these particles could be selected with powerful magnets. Neutral particles, and in particular the $\gamma$s produced in the decay of the abundantly produced $\pi^0$s, were of course not affected by the

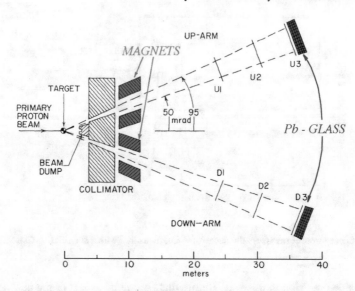

**Fig. 1.3** The magnetic spectrometer of the Columbia/FNAL experiment E-70. From: Appel, J.A. et al. (1975). *Nucl. Instr. and Meth.* **127**, 495

magnets and did not produce any signals in the wire chambers. They were detected in arrays of crystals installed at the end of both spectrometer arms.[1]

The crystals used in this experiment were not based on the production of scintillation light. Instead, the signals they produced were the result of the Čerenkov effect, which leads to the production of light by particles that travel faster than the speed of light in the medium they traverse, lead glass in this case. Details of this mechanism are described in the next subsection. The main advantage of these lead glass crystals was the cost, which is only a small fraction of the cost of scintillating crystals.

Whereas these crystals turned out to be excellent detectors for electrons and $\gamma$ rays in an energy domain that was several orders of magnitude beyond that for which they were initially intended, attempts to detect also other particles, such as protons and charged pions, led to disappointing results. After the first successful use of NaI(Tl) crystals for detecting GeV-type electrons and $\gamma$s by Hofstadter and his colleagues at SLAC in the late 1960s [3], it was believed that hadrons could be detected with the same type of precision, provided the detectors were made sufficiently large. To test this idea, the researchers collected as many NaI(Tl) crystals as they could get their hands on and assembled these to a calorimeter with a total mass of about 450 kg (0.12 m$^3$). They exposed this device to particle beams (predominantly pions, with small admixtures of electrons and muons of the same energy) of different energies, ranging from 4–16 GeV. Figure 1.4 shows the signal distribution for 8 GeV $\pi^-$ measured with this detector [4].

---

[1]The primary experimental goal of this experiment was the detection of high-energy electrons and positrons produced in the target. These studies led in 1977 to the discovery of the $b$-quark [2].

**Fig. 1.4** Pulse height
distribution for 8 GeV
negatively charged particles
in a 450 kg array of NaI(Tl)
crystals. From: Hughes,
E.B. et al. (1969). *Nucl. Instr.
and Meth.* **75**, 130

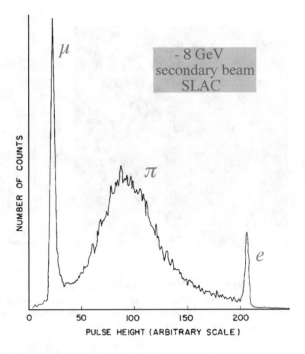

They interpreted the observation that the signals from the pions were, on average, only half as large as those from the electrons as evidence that about half of the energy leaked out of this detector. The fact that Monte Carlo simulations indicated that this leakage was actually much smaller was blamed on a flaw in the simulations. The results obtained at other energies (4, 12 and 16 GeV) were very similar to those shown in the figure: pion signals that were typically only half as large as those of electrons of the same energy, an asymmetric hadronic response function, and a hadronic energy resolution ($\sigma/E$) that was approximately *independent* of the energy. Their conclusion from these observations was that the calorimeter was too small to perform well for hadron detection. However, tests of a huge, 60-ton homogeneous calorimeter consisting of liquid scintillator built a few years later at Fermilab revealed that shower leakage was not the dominating problem for the poor hadronic performance. Tests of that detector showed that pion signals were also substantially smaller than electron ones for the same energy, and that the hadronic energy resolution essentially did not improve with increasing energy [5].

We now know that the basic reason for the different hadronic and em calorimeter responses lies in the fact that in the absorption of hadronic showers, a significant fraction of the energy carried by the showering particle is *invisible*, i.e., it does not contribute to the calorimeter signal. The main source of invisible energy is the energy used to release nucleons from nuclei, including the nuclear recoil energy. Additional, smaller contributions come from neutrinos and muons (mainly from $\pi$

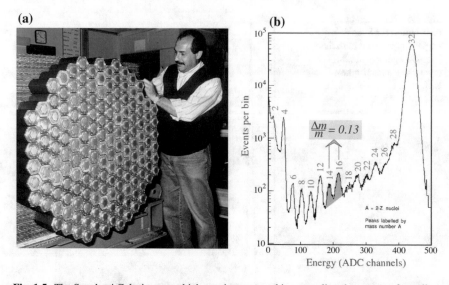

**Fig. 1.5** The Spaghetti Calorimeter, which was instrumental in unraveling the secrets of excellent hadronic calorimeter performance (**a**). Photograph courtesy CERN. Example of the experimental performance that can be obtained with an excellent hadronic calorimeter (**b**). From: Young, G.R. et al. (1989). *Nucl. Instr. and Meth.* **A279**, 503

and $K$ decay in flight) escaping the detector, and from signal saturation for densely ionizing particles. These effects are discussed in detail in Chap. 2.

Once the reasons for the poor performance of hadron calorimeters were understood, dedicated efforts could be undertaken to improve that performance. These efforts, which are described in detail in Chap. 8, were very successful. An example of a calorimeter developed in the context of this R&D work is depicted in Fig. 1.5a. This so-called *Spaghetti Calorimeter* was for 25 years holder of the best hadronic performance characteristics. A special feature of this calorimeter was that it was *compensating*, i.e., it produced signals for electrons and hadrons of the same energy that were, on average, equally large (unlike the crystals from Fig. 1.4). An example of results that were obtained with a compensating calorimeter is shown in Fig. 1.5b. This is a signal distribution from the early days of heavy-ion acceleration with the CERN SPS. The beam was obtained by accelerating $^{32}$S ions to an energy of 200 GeV per nucleon. However, it turned out that the beam contained "satellites" consisting of ions with the same charge/mass ratio, such as $^4$He, $^{12}$C, $^{16}$O, etc. [6]. This was discovered by dumping the beam of the accelerated ions into a (very good) calorimeter. The satellites showed up as peaks with intensities at the level of $10^{-3}$ in the spectrum. Note that the resolution in this plot is very similar to that of the $\gamma$ ray spectrum shown in Fig. 1.2.

In recent years, even better results than those obtained with compensating calorimeters have become possible thanks to the use of the so-called *dual-readout* technique. This technique and the possibilities it offers are the topic of Sect. 8.3.

Calorimeters such as the one shown in Fig. 1.5a differ from the crystals in the sense that they are *sampling* devices. The functions of particle absorption and signal generation are in these detectors performed by different materials, e.g., lead and plastic scintillator in this example. The ratio between the passive material (lead) and the active one (scintillator) determines the *sampling fraction*, i.e., the fraction of the energy of the absorbed particle that is deposited in the active material and thus contributes to the calorimeter signal. In this calorimeter, that fraction is 2.4%. An important reason why almost all calorimeters used nowadays in particle physics experiments are of the sampling type is cost, since the signal generating materials are typically much cheaper than the absorber material. Also the fact that the density of the absorber material can be much larger is an important consideration, especially in $4\pi$ experiments, since it makes it possible to construct much more compact detectors.

Yet, the small sampling fraction does not necessarily imply that the energy resolution of sampling calorimeters is worse than for *homogeneous* ones, i.e., calorimeters such as crystals where the functions of particle absorption and signal generation are performed by the same material. As a matter of fact, sampling calorimeters such as the one shown in Fig. 1.5a measured hadrons with an energy resolution that was a factor of three better than for the best homogeneous calorimeter [5, 7].

However, the good energy resolution that can be obtained for high-energy particles was not the only, and not even the main, reason why calorimeters have become the most important components of most modern experiments in particle physics, and definitely in experiments carried out at high-energy particle colliders such as the Large Hadron Collider at CERN. It turned out that calorimeters are ideal instruments for providing information that is crucial in the context of the physics to be studied in such experiments and, especially, that they can provide this information very fast.

Figure 1.6 shows a picture of one of the first large calorimeters that operated in an accelerator based experiment. The experiment, WA1, was intended to study interactions induced by neutrinos produced by the Super Proton Synchrotron at CERN. This detector had an instrumented mass of about 1 kiloton. It consisted of large slabs of iron, interleaved with active material that generated the signals. In the upstream part of the detector (the first 6 m), this active material consisted of scintillating plastic. There were in total 105 such sampling layers. This part of the detector served as the neutrino target. In the rear part of the detector, the iron slabs were thicker (15 cm instead of 5 cm), and they were interleaved with wire chambers, which allowed the localization of passing charged particles with good precision.

By comparing to total signal registered in all scintillator plates combined with preset thresholds, the occurrence of neutrino interactions in the target section could be determined in real time. The rear part of the detector made it possible to determine the trajectory of muons in case these were produced in the interactions. Since the iron was also magnetized, the momenta of such muons could be measured as well. The detection of muons was a crucial aspect of this experiment, since these particles distinguished charged current neutrino reactions from neutral current ones. Moreover, dimuon events were a signature of the production of charmed particles. These were hot topics in the 1970s, and this experiment made major contributions in this

**Fig. 1.6** The WA1 neutrino detector combination that operated at CERN from 1976 to 1984. Large slabs of absorber material (iron) are interleaved with layers of a plastic scintillator. The rear part of the detector, located at the left-hand side of the picture, is instrumented with wire chambers intended for tracking muons generated in charged current interactions and/or charmed particle production. Photograph courtesy CERN

domain, as well as to the understanding of details of the nucleon structure, for which (anti-)neutrinos are very important tools.

The role of calorimeters became even more important with the advent of colliding-beam accelerators. A crucial aspect of experiments at such colliders is *hermetic coverage*, i.e., the detectors have to cover as large a fraction as possible of the $4\pi$ steradians surrounding the interaction point of the colliding particles. If that is achieved, the detectors can provide crucial information on the *energy flow* in the events. For example, a comparison between the total signals measured in any two hemispheres gives information about the *missing transverse energy*. This may indicate the production of an energetic neutrino, which does not leave a trace in the calorimeter. The calorimeter data can also be used to identify other types of particles produced in the interactions. For example, the energy deposit profile and/or the time structure of the signals are important tools to recognize electrons, and the fact that a charged particle penetrates the calorimeter indicates that it is a muon.

The total measured calorimeter signal is a measure for the centrality of the collision (deep inelastic scattering), and the concentration of energy in certain regions of the calorimeter ("jets") is a signature for a fragmenting quark of gluon produced in the interactions. Since the rate of the particle interactions in such colliders may be extremely high (e.g., $\mathcal{O}(\text{GHz})$ in the Large Hadron Collider), the calorimeter is also crucial for determining which of these events should be retained for further, offline,

**Fig. 1.7** The UA2 calorimeter that operated at CERN's proton–anti-proton collider ($Sp\bar{p}S$) from 1980 to 1986 and played a crucial role in the discovery of the $W$ and $Z$ bosons [8]. Photograph courtesy CERN

inspection. This triggering capability is possibly the most important reason for the dominant role of calorimeters in modern collider experiments.

In 1982, the UA1 and UA2 (see Fig. 1.7) experiments discovered the $W$ intermediate vector boson [9, 10] on the basis of the signatures mentioned above: an energetic charged lepton (electron, muon), in combination with missing transverse energy (caused by the neutrino accompanying the charged lepton in the decay $W \rightarrow l\nu$. This discovery contributed in no small measure to the way in which experiments are designed nowadays.

A relatively recent development is that of *imaging calorimeters*. These are based on liquid-argon technology, instrumented as Time Projection Chambers, and produce event images that rival those from bubble chambers (Fig. 1.8). These devices are typically used in neutrino oscillation experiments, located at different distances from the source [11]. For example, neutrinos produced at Fermilab are (scheduled to be) detected at distances varying from a few km ($\mu$BOONE) to 1,300 km (DUNE) [12]. Whereas measuring the energy carried by these particles is an important goal of such experiments, they typically emphasize complete imaging of the detected events.

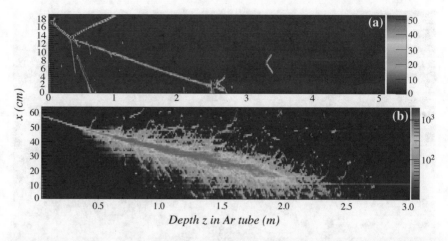

**Fig. 1.8** Examples of cosmic ray events in a large liquid argon TPC. Diagram **a** shows an interaction by a charged particle, in which three charged particles are produced, plus possibly some neutral objects. Several $\delta$-rays are also visible along the track of the particle that re-interacts after about 2 m. Diagram **b** shows a textbook example of a developing em shower in argon. From: Ereditato, A. et al. (2013). *J. Instrumentation* **8**, P07002

## 1.3  Types of Calorimeters Used as Particle Detectors

As indicated in the previous subsection, calorimeters are either homogeneous or have a sampling structure. In the latter case, the functions of particle absorption and signal generation are exercised by different materials, called the passive and active media, respectively. In homogeneous calorimeters, the entire volume is active. This category includes detectors consisting of crystals which either produce scintillation or Čerenkov light. Figure 1.9 shows a picture of a sector of the electromagnetic (em) calorimeter of the CMS experiment at CERN's Large Hadron Collider. This detector consists of large numbers of $PbWO_4$ crystals [13], which are the most dense crystals that produce scintillation light. Other scintillation crystals that are used in particle physics experiments include BGO [14], CsI(Tl) [15, 16]. The most commonly used detectors that produce Čerenkov light are made of lead glass [17], with $PbF_2$ as a more dense, but also more expensive alternative [18]. For experiments in the energy domain below 10 MeV, calorimeters based on semiconductors are most frequently used. For example, in nuclear $\gamma$ ray spectroscopy, high-purity germanium detectors are the standard, while silicon detectors are commonly used for $X$ ray measurements in astronomy and elsewhere. The energy resolutions obtained with these devices are considerably better than the ones shown in Fig. 1.2 for scintillating crystals [19].

Another category of homogeneous calorimeters is based on liquified noble gases, which are also known to be very bright scintillators, albeit in the far ultraviolet wavelength region, which complicates their use [20]. One experiment that uses this feature is MEG, which searches for the extremely rare decay $\mu^+ \rightarrow e^+\gamma$ at PSI in Villigen (Switzerland) and uses a 900 L liquid xenon [21] detector for this purpose

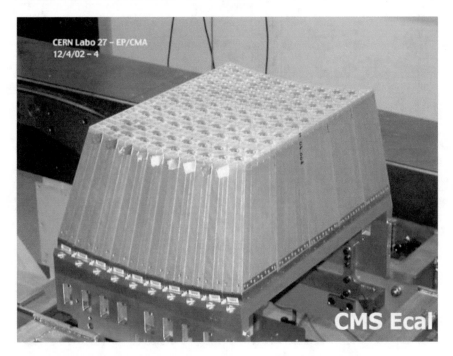

**Fig. 1.9**  A sector of the CMS ECAL, which consists of scintillating PbWO$_4$ crystals. Photograph courtesy CERN

[22]. Liquid argon is used on a large scale for imaging calorimeters (Fig. 1.8), but in this case the ionization charge produced in the particle absorption process is the source of the signals.

By far the largest homogeneous calorimeters in use are based on the production of Čerenkov light in water. SuperKamiokande (Fig. 1.10) operates a detector containing 55,000 ton of ultrapure water in a Japanese mine [23]. Even larger instruments use the natural environment for particle detection. For example, IceCube [24] has instrumented 1 km$^3$ of Antarctic ice, and several other efforts are under way to do the same in the Mediterranean [25]. The Auger experiment [26] uses (scintillation) light production in a very large volume of the Earth's atmosphere to measure the highest energy cosmic rays that reach us from outer space.

The calorimeter systems used in experiments at the highest energy particle colliders are typically of the sampling type, mainly because of cost and compactness considerations, although crystals are still popular at $e^+e^-$ rings, especially those operating at the $J/\Psi$ [16] and $\Upsilon$ [27] resonances. The active material in sampling calorimeters generates signals in the form of (scintillation or Čerenkov) light or electric charge, by direct ionization or the production of electron-hole pairs (semi-conductors). The absorber material is typically lead for the detection of em particles, and iron or copper for hadrons. In some cases, depleted uranium has been used, based on the (incorrect) assumption that this would provide unique advantages.

**Fig. 1.10** Picture taken inside the Super-Kamiokande detector, showing part of the 11,200 PMTs that detect the Čerenkov light when the vessel is filled with 55 kilotons of ultrapure water. Note the three technicians standing on the left.  Photograph courtesy Super-Kamiokande

The structure of these sampling calorimeters has evolved over time. Initially, it was assumed that the active and passive layers had to be arranged in the from of a "sandwich" structure, with the layers oriented perpendicular to the travel direction of the envisaged particles (Fig. 1.11a). However, in the 1980s it was discovered that there was absolutely no fundamental reason that made this necessary, and that other arrangements could offer major practical advantages, while barely affecting the calorimeter performance. In many modern calorimeters, the active layers are therefore distributed in very different ways inside the absorber structure. For example, in the ATLAS hadron calorimeter, tiles of plastic scintillator are oriented *along* the direction of the incoming particles (Fig. 1.11b). And in the CMS HF calorimeter, large numbers of thin quartz fibers that run in the same direction as the incoming particles collect Čerenkov light produced in the absorption process and transport it to the rear end of the calorimeter (Fig. 1.11c). In the ATLAS em calorimeters, the active medium (liquid argon in this case) is contained in a sort of "accordion-like" structure of containers inside the lead absorber structure (Fig. 1.12).

The reason for these particular choices lies in the fact that the mentioned calorimeters have to operate in a $4\pi$ environment. The chosen structure offers some clear advantages compared with the "sandwich" one. For example,

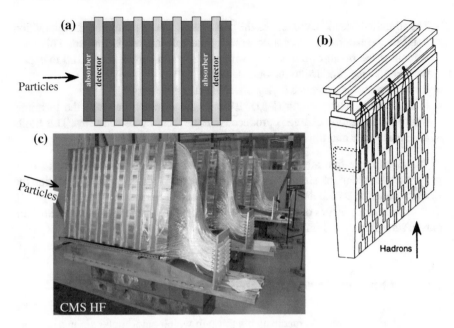

**Fig. 1.11** Different sampling structures used in calorimeters. The "sandwich" structure, used in the ZEUS calorimeters (**a**). The "tile" structure, used in the ATLAS hadron calorimeter (**b**). Image courtesy CERN. The "spaghetti" structure, used in the very forward region of CMS (**c**). Photograph courtesy CERN

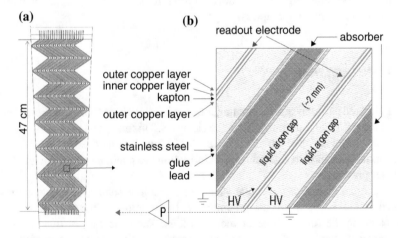

**Fig. 1.12** The "accordion" structure used in the ATLAS em calorimeter. From: Aad, G. et al. (2010). *Eur. Phys. J.* **C70**, 755

- The so-called "dead" volume, i.e., the part of the space surrounding the interaction point of the beams that is not covered by the calorimeter, is limited. This dead volume is the result of the need to support the multi-ton detectors and to export the signals they produce to the outside world.
- It makes it easier to realize a projective detector geometry
- In the case of liquid argon (LAr), It limits the distance between the locations where the ionization charge is produced and the front-end electronics. This limits the capacitance and thus the electronic noise of the system.

In calorimeters that are not part of a $4\pi$ detector system, there is still a strong preference for the much simpler sandwich structure. As an example, we mention the MINOS detector, which detects neutrinos produced at Fermilab, 800 km away [28]. This calorimeter consists of 2.54 cm thick iron absorber plates interleaved with 1 cm plastic scintillator (cf. Fig. 1.6).

## 1.4    Techniques Used for Signal Generation in Calorimetry

In this section, the various mechanisms through which calorimeter signals are generated are described. The discussion is limited to the general aspects that make these mechanisms useful for application as active media for calorimetric particle detection. Specific features, such as signal quenching effects for densely ionizing particles or non-linearities caused by active media operating in the Geiger mode, are elaborated upon in the relevant later chapters.

### 1.4.1   Scintillation

When charged particles traverse matter, they lose energy through the electromagnetic interaction with the Coulomb fields of the electrons. This energy may be used to ionize the atoms or molecules of which the traversed medium is composed, or to bring these atoms/molecules into an excited state. Scintillation is a phenomenon associated with the latter process.

The excited atomic or molecular states are unstable. Usually, the excited atom or molecule quickly returns to the ground state. In this process, the excitation energy is released in the form of one or more photons. The timescale of this process is determined by the excitation energy, by the number of available return paths, and by the quantum numbers of the states involved (wavefunction overlap). When the energy differences are such that the emitted photons are in the visible domain, this process is called *fluorescence* or *scintillation*. Typical timescales range in that case from $10^{-12}$ to $10^{-6}$ s, although exceptions in either direction may occur. In general, the timescales get shorter as the molecules get more complex. This can be simply understood from the fact that the density of excited states, and therefore the number of different ways in which an excited molecule can get rid of its excess energy increases sharply with the complexity of the molecule.

Relatively simple scintillating crystals, such as NaI(Tl) and BGO, have decay times of several hundred ns, orders of magnitude longer than the decay times of complex organic scintillators, such as those based on anthracene or polystyrene.

Historically, scintillation was the first physics process to be used for the generation of calorimetric signals. And until this day, a large number of calorimeters in a wide variety of particle physics experiments rely upon scintillation light as the prime source of information. Two inventions have played a crucial role in the development of scintillator-based particle detectors in general, and calorimeters in particular:

- *The photomultiplier tube.* Almost 70 years old, the PMT which allows the conversion of single photons into electric signals, is still playing a crucial role in many experiments. The sensitivity to single photons and the essentially noise-free amplification offered by these devices are very attractive features [29]. Although some new devices based on semiconductor applications (such as the Hybrid Photo Detector [30], the Avalanche Photo Diode [31] and especially the silicon photomultiplier [32, 33]) have more or less successfully addressed some PMT weaknesses, such as the sensitivity to external magnetic fields, the essentially noise-free signal amplification offered by PMTs is still an extremely attractive feature.
- *Wavelength shifters.* These devices absorb the scintillation light and re-emit it at a lower energy (longer wavelength). This development made it possible to apply scintillator-based calorimeters in experiments requiring hermetic coverage, such as the $4\pi$ experiments in a colliding-beam setup. The light produced in scintillator plates oriented perpendicular to the direction of the incoming particles can be wavelength-shifted and at the same time redirected towards the rear end of the calorimeters, where it can be converted into electric signals. Figure 1.13 shows schematically the readout of scintillator calorimeters with and without wavelength-shifting plates. The price to be paid for these advantages is a loss of light, because of inefficiencies in the process and a longer signal duration, since the wavelength shifters are usually somewhat slower than the scintillators whose light they shift.

The development of plastic optical fibers has also greatly influenced the design of scintillator calorimeters, especially after it became clear that there is no reason why active material has to be oriented in a particular direction. Scintillating fibers usually consist of a polystyrene core (index of refraction $n = 1.59$), surrounded by one or several layers of cladding with (gradually) lower values of $n$.

Unlike the optical fibers used for telecommunication purposes, which are designed to *transport light injected along the fiber axis*, the scintillating fibers used in particle physics experiments are both the *source of the light* (generated isotropically) and the medium through which this light is transported to a place where it can be converted into an electric signal. The fraction of the light that is trapped is proportional to the *numerical aperture* $\sqrt{n_{core}^2 - n_{clad}^2}$, and most of the light is traveling near the *critical angle*, defined as $\theta_{cr} = \arcsin(n_{clad}/n_{core})$ [35].

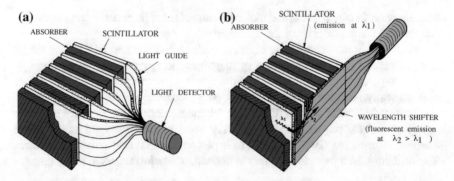

**Fig. 1.13** Schematic of the readout systems of scintillator calorimeters without (**a**) and with (**b**) wavelength-shifting plates [34]

Apart from these chemically doped optical fibers, undoped plastic fibers are also being applied in particle detectors. For example, clear plastic fibers based on a PMMA core ($n = 1.49$) surrounded by lower-index (fluorinated plastic) cladding material is used to detect Čerenkov light produced in the particle absorption process (see Sect. 1.4.2).

Optical fibers are being used in many calorimeters, either as the active medium sampling the showers, or as wavelength shifters, converting the scintillation light, e.g., blue light from scintillator plates, to a longer wavelength (e.g., green) and transporting it to light detectors located in a convenient position.

Among the advantages offered by such fibers, we mention

- The perfectly hermetic calorimeter structure that can be achieved,
- The very high sampling frequency (good energy and position resolution!) that can be obtained using fibers as the active medium,
- The high signal speed that can be obtained,
- The arbitrary granularity allowed by the fiber structure,
- The high light yield that can be achieved, and
- The excellent cost/performance ratio

We elaborate on the relevance of these factors, and on the specific advantages of plastic as active material for *hadron* calorimeters in the next chapters.

## 1.4.2 Čerenkov Radiation

When a charged particle travels faster than the speed of light in a certain medium ($v > c/n$, or $\beta = v/c > 1/n$, with $n$ the medium's index of refraction), it loses energy by emitting Čerenkov radiation. This radiation is emitted at a characteristic angle, the Čerenkov angle $\theta_C = \arccos (n\beta)^{-1}$, with the direction of the particle. Therefore,

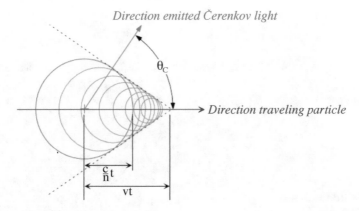

**Fig. 1.14** The principle of Čerenkov light emission by a superluminal particle. In a time $t$, the particle travels a distance $vt$, while the light it emits travels a distance $ct/n$. The wavefronts of the light emitted by such a particle form a cone with half-opening angle $\theta_C$

this radiation forms a cone with half-opening angle $\theta_C$ (Fig. 1.14). The amount of energy is proportional to $\sin^2 \theta_C$ [29, 36].

The spectrum of this Čerenkov radiation exhibits a characteristic $1/\lambda^2$ dependence and, therefore, the visible part of the Čerenkov spectrum is experienced as blue light. This blue light can be abundantly observed in highly radioactive environments, e.g., the moderating liquids in nuclear reactors. It is also a source of light deep in the oceans, where it is created by penetrating cosmic rays.

The emission of Čerenkov light is only a very minor source contributing to the energy loss of the particles. For example, in water, a charged particle with $\beta \simeq 1$ loses about 400 eV/cm in the form of visible Čerenkov photons. That is some four orders of magnitude less than its energy loss through other processes, in particular ionization (2 MeV/cm).

Since the Čerenkov mechanism is sensitive to the *velocity* of particles, it can be used to determine the *mass* of particles of which the *momentum* has been determined by means of deflection in a magnetic field. A variety of devices (threshold Čerenkov counters, differential Čerenkov counters, ring imaging Čerenkov detectors) have been developed to separate electrons, pions, kaons, protons and deuterons from each other, exploiting this effect.

As we will see in later chapters, calorimeters based on the detection of Čerenkov light exhibit some interesting properties, which may be ideal for certain very specific applications, e.g., jet detection very close to the beam pipe in LHC experiments, or dual-readout calorimetry (Sect. 8.3).

A very important aspect of Čerenkov light is its *instantaneous* character. There are no delaying factors, such as the lifetime of a metastable excited state, which affect the time characteristics of detectors based on scintillation light. Therefore, Čerenkov detectors, including calorimeters, are the instruments of choice for experiments in which ultimate signal speed is required.

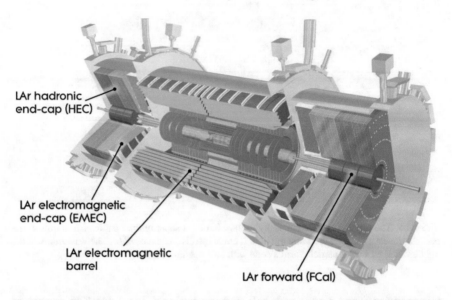

LAr hadronic
end-cap (HEC)

LAr electromagnetic
end-cap (EMEC)

LAr electromagnetic
barrel

LAr forward (FCal)

**Fig. 1.15** Schematic drawing of the ATLAS LAr calorimeter system. In the middle, the barrel cryostat hosts two electromagnetic wheels, at each end the end cap cryostats host two concentric electromagnetic wheels, two hadronic wheels and three forward calorimeter wheels. Image courtesy CERN

### 1.4.3 Ionization

When charged particles traverse matter, they may ionize the atoms of which this matter consists. One or several electrons are released from their Coulomb field in this process, leaving behind an ionized atom. Collection of these liberated electrons is applied as the signal-producing technique in a wide variety of particle detectors. The electrons produced along the trajectory of the ionizing particle may or may not be amplified in this process.

In ionization chambers based on liquid media, no amplification takes place. A potential difference applied over the gap containing the liquid separates the electrons from the ions. The electrons are collected at the anode, the ions at the cathode. In order for this method to work properly, the mean free path of the electrons in the liquid should be long, considerably longer than the size of the gap between the electrodes. Therefore, noble liquids such as argon, krypton and xenon, which have no desire to capture loose electrons wandering around since all the electronic shells of their atoms are filled, are the media of choice in these detectors. The fact that the use of such liquids requires cryogenic operating conditions typically represents a major experimental challenge. Especially, hermetic coverage in a $4\pi$ environment is very hard to achieve (Fig. 1.15).

To ensure a sufficiently long mean free drift path for the electrons, very stringent purity standards have to be met in noble-liquid ionization chambers. In particular, contamination by electro-negative elements such as oxygen have to be kept below the 1 ppm ($10^{-6}$) level. Calorimeters based on noble liquids as active media have been used in particle physics experiments since the 1970s. The technique was pioneered with liquid argon (LAr). Liquid argon is cheap, abundantly available and the required purity levels can be easily achieved and maintained. Among the largest LAr calorimeter systems operating today, we mention the ATLAS experiment at CERN's Large Hadron Collider. Previously, the D0 experiment at Fermilab's Tevatron collider and the H1 experiment at the HERA electron–proton collider (DESY, Hamburg) were centered around large LAr sampling calorimeters.

A very large *homogeneous* LAr system operating as a Time Projection Chamber (cf. Fig. 1.8) was pioneered for ICARUS [37] in the Gran Sasso laboratory. A 170 kiloton successor ($\mu$BooNE) detects neutrino interactions at Fermilab and even much larger devices (DUNE) are being planned for the future.

Other noble liquids, krypton (LKr) and xenon (LXe) are much more expensive and are therefore only used in applications requiring the specific advantages offered by these liquids, such as a higher density, or a higher $Z$ value. For example, the electromagnetic sampling calorimeter of the NA48 experiment at CERN, which studied $CP$ violation in the $K^0$ system, and its successor NA62, intended for the study of extremely rare kaon decays, is based on LKr as the active medium. This makes it possible to achieve the energy resolution for $\gamma$ detection required by the sensitivity goals of these experiments [38].

Apart from the ionization charge produced by charged particles traversing the detector, LXe is also a very bright and fast scintillator. This feature, combined with its high $Z$ value (54) make it an excellent detector for $\gamma$ rays. Several dedicated experiments make use this feature, e.g., the MEG experiment mentioned in Sect. 1.3. The properties of liquid xenon also make it a detector of choice for certain dark matter searches and neutrinoless $\beta\beta$ decay. Elastic scattering of weakly interacting massive particles (WIMPs) gives the heavy Xe nucleus a relatively large recoil energy, which may produce a measurable signal because of the high light yield combined with the sensitivity for ionization charge. And since xenon consists for 8.9% of the isotope $^{136}$Xe, which can only decay through the conversion of two neutrons into two protons, LXe can at the same time act as the source *and* the detector for this rare process. Several experiments based on large LXe TPCs are currently operating or being prepared with these goals in mind [39].

Unlike scintillating crystals, noble liquids are very radiation hard. They can also be rather easily replaced if needed. This has played a role in the choice of a lead/LAr sampling calorimeter for the ATLAS experiment.

A totally different class of ionization calorimeters is based on gaseous media. In these devices, the electrons produced in the ionization process undergo considerable multiplication before being collected at the anode. As they are accelerated in the electric field between the anode and cathode, the electrons may acquire enough energy to ionize other atoms and thus release *secondary* electrons. These may in turn release tertiary electrons, etc. The result is an *avalanche* of electrons arriving at the anode and constituting the signal. Since electric fields in the vicinity of a charged object (the anode) strongly depend on the distance $r$ from that charged object, and since the energy acquired by the electrons is proportional to the electric field strength, this multiplication process works best in the immediate vicinity of the anode (small values of $r$). For this reason, the anode is often made of very thin (30 μm) wires.

Wire chambers may operate in a variety of modes, depending on the type of gas mixture and on the voltage difference between anode and cathode: the proportional mode, the streamer mode, the Geiger mode, etc. The time needed for the charge to arrive at the anode may provide information about the spatial coordinates of the particle that caused the signals. This principle is applied in *drift chambers* and in *Time Projection Chambers* [40, 41]. A large number of calorimeter systems rely on wire chambers or tubes of some sort to provide the experimental signals. Especially when very large surface areas have to be covered, this is often the most cost effective solution available.

Thin wires make such detectors very fragile. A short circuit caused by a broken wire may wipe out a large section of the detector system, as has been experienced by several experiments. In the past 25 years, we have witnessed an enormous development of detectors in which the strong electric fields needed for the multiplication of the ionization charge are created in some other way than by means of thin wires. The general term for such devices is "micropattern gas detectors." Thin gaps, tiny holes or some micromesh structure provide the field-shaping geometry, and new developments in photolithography and microelectronics have led to a new class of extraordinary detectors [41]. Not surprisingly, these devices have become increasingly popular with designers of tracking systems for new experiments.

Finally, there are also solid state devices that are being used as detectors of charges produced by passing ionizing particles. Silicon, germanium and gallium arsenide have all been applied for particle detection since more than 50 years. The outer-shell atomic levels of these semiconductor crystals exhibit a band structure, consisting of a valence band and a conduction band, separated by a "forbidden" energy gap, in which no energy levels are available. An ionizing particle passing through such a semiconductor excites electrons from the valence band into the conduction band. For every electron that jumps into the conduction band, a *hole* remains in the valence band. This hole is positive relative to the sea of negative electrons in the valence band. Therefore, it acts as a positive charge carrier, and its movement through the semiconductor crystal constitutes an electric current, just as does the movement of the electrons in the conduction band. The electron–hole pairs created by ionizing particles may be collected by means of an electric field.

This technique has several advantages. The energy gap between the valence and conduction bands is very narrow, typically of the order of 1 eV and, therefore, very little energy is required for the production of one electron–hole pair. For example, in silicon every 3.6 eV of deposited radiation energy yields one electron–hole pair. This is typically one order of magnitude less than the energy needed to produce one electron–ion pair in gases and two orders of magnitude less than the energy required for the production of one photoelectron in scintillation counters. Therefore, semiconductor crystals offer the potential of excellent energy resolution in detectors in which fluctuations in the number of primary charge carriers are the limiting factor for resolution.

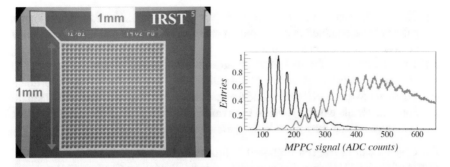

**Fig. 1.16** Example of a silicon photomultiplier, and the spectra it produces when exposed to two different sources of visible light [34]

Charge amplification can also be applied in semiconductors. Silicon is an excellent detector of visible light, especially in the longer wavelength region where the quantum efficiency of photocathodes used in PMTs is decreasing. Silicon diodes, which are a common tool in movement detectors, make use of this characteristic. The photoelectrons may also be internally amplified, which is the operating principle of Avalanche Photo Diodes. If the applied voltage is sufficiently large, a detected photon may cause the device to discharge, and thus act like a Geiger counter. Silicon photomultipliers (SiPM) consist of large numbers of extremely small silicon pixels, each of which operates in that way [32]. Figure 1.16 shows an example of a SiPM that contains 625 pixels on a surface area of only 1 mm$^2$. As illustrated by the right diagram, these detectors are excellent single-photon counters. The number of photons constituting a given signal can be precisely determined for up to 20 photons or so. In that sense, as well as in their capability to operate in a magnetic field, they present a clear advantage over PMTs. An additional advantage of semiconductor crystals concerns their response time. Because of their greater density and compact structure, they may be considerably faster than other detectors based on the collection of ionization charge.

## 1.4.4 Cryogenic Phenomena

There is a class of highly specialized detectors that employ calorimetric methods to study a series of very specific phenomena in the boundary area between particle physics and astrophysics: dark matter, solar neutrinos, magnetic monopoles, nuclear double $\beta$-decay, etc. All these issues require precise measurements of small energy deposits. In order to achieve that goal, the mentioned detectors exploit phenomena that play a role at temperatures close to zero, in the few milli-Kelvin to 1 Kelvin range. These phenomena include:

($a$) Some elementary excitations require very little energy. For example, Cooper pairs in superconductors have binding energies in the $\mu$eV–meV range and may be broken by phonon absorption.

($b$) The specific heat for dielectric crystals and for superconductors decreases to very small values at these low temperatures.

($c$) Thermal noise in the detectors and the associated electronics becomes very small.

($d$) Some materials exhibit specific behavior (e.g., change in magnetization, latent heat release) that may provide detector signals.

The devices that have been proposed in this context are typically still in the early phases of the R&D process. In many cases, this R&D involves fundamental research in solid-state physics and materials science. However, some devices have reached the stage where practical applications have been successfully demonstrated. Among these, we mention

- *Bolometers*, which are based on principle ($b$). These are calorimeters in the true sense of the word, since the energy deposit of particles (in an insulating crystal at very low temperature) is measured with a resistive thermometer.
- *Superconducting Tunnel Junctions*, in which the quasi-particles and -holes (Cooper pairs) excited by incident radiation tunnel through a thin layer separating two superconducting materials.
- *Superheated Superconducting Granules*, which are based on the fact that certain type I superconductors can exhibit metastable states, in which the material remains superconducting in external magnetic fields exceeding the critical field. These detectors are usually prepared as a colloid of small (diameter 1–100 $\mu$m) metallic granules suspended in a dielectric matrix (e.g., paraffin). Heat deposited by an interacting particle may drive one of several granules from the superconducting to the normal state. The resulting change in magnetic flux (disappearance of the Meissner effect) may be recorded by a pickup coil.

Figure 1.17a shows the operating principle of a typical cryogenic calorimeter. It consists of an absorber with heat capacity $C$, a thermometer and a thermal link with a heat conductance $g$ to a reservoir with temperature $T_B$ [46]. The thermometer is typically a thin superconducting strip that operates very close to the transition temperature between the superconducting and normal phases. A small local increase in the temperature may dramatically increase the electric resistance, and thus lower the current flowing through this circuit. The temperature increase needed for this to happen may be caused by phonons created by particles interacting in the absorber, which travel to the surface and break Cooper pairs. The extremely low temperature at which these detectors are operating is needed to reduce the thermal noise, which limits the size of the measurable signals. The use of superconductors as cryogenic particle detectors is motivated by the very small binding energy of the Cooper pairs, $\sim$1 meV, compared to 3.6 eV needed to create an electron–hole pair in silicon. Thus, compared to a semiconductor, several orders of magnitude more free charges are produced, which leads to a much higher intrinsic energy resolution (Fig. 1.17b).

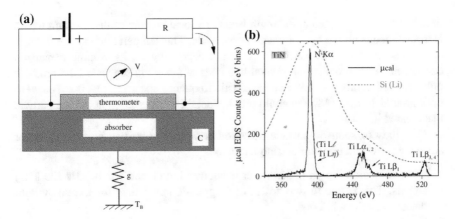

**Fig. 1.17** The principle on which a cryogenic calorimeter is based (**a**). Spectrum of X-rays of titanium nitrate, measured with a cryogenic calorimeter, and with a standard Si(Li) semiconductor detector (**b**). From: Pretzl, K. (2000). *Nucl. Instr. and Meth.* **A454** 114

Because of their sensitivity to very small energy deposits, cryogenic calorimeters are widely used in the search for dark matter, and in particular for low-mass WIMPs. As a matter of fact, the most stringent limits on WIMPs with masses less than 5 GeV/$c^2$ come from cryogenic experiments such as CDMS [42], EDELWEISS [43] and CRESST [44], which all operate bolometric detectors with masses in the 10 kg range at temperatures well below 1 K, in tunnels or deep mines. Plans for upgrades to 100 kg or more exist in all cases. An example of a cryogenic calorimeter looking for neutrinoless $\beta\beta$ decay is CUORE [45], which is building a detector containing 740 kg worth of $TeO_2$ crystals (i.e., 240 kg of the isotope of interest, $^{130}Te$), which will operate at a temperature of 15 mK in the Gran Sasso Laboratory. Superconducting Tunnel Junctions have found useful applications in astronomy, where they are used to detect radiation in the sub-mm wavelength domain.

There is a considerable amount of effort going into the development of these and many related, similarly ingenious devices. However, this highly specialized work falls somewhat outside the scope of this book. The interested reader is referred to reviews of this field that can be found in [46, 47]. A recent review on dark matter searches is given in [39].

## 1.4.5 Acoustic Signals

At the other end of the energy scale covered by particle calorimeters, i.e., in the Joule domain ($>10^{18}$ eV), attempts are being made to use acoustic signals, especially for the detection of cosmic neutrinos with ultra-high energies. The pressure signals are produced by the particle showers that evolve when neutrinos interact with nuclei in water. The resulting energy deposition in a cylindrical volume of a few centimeters in radius and several meters in length leads to a local heating of the medium which is instantaneous on the hydrodynamic time scale. This temperature change induces an expansion or contraction of the medium depending on its volume expansion coefficient. After propagating several hundreds of meters in sea water, the pressure pulse created in this way has a characteristic frequency spectrum that is expected to peak around 10 kHz [48] and could provide detailed information about the particle that created it.

The study of acoustic particle detection is motivated by two potentially major advantages over an optical neutrino telescope:

1. The attenuation length in sea water is about 5 km (1 km) for 10 kHz (20 kHz) signals. This is one to two orders of magnitude larger than for Čerenkov light detected by the telescope.
2. The sensors can be more compact and the readout electronics simpler for acoustic measurements.

In principle, this offers the promise of a much simpler telescope covering a much larger fiducial volume. Potential disadvantages include:

1. The speed of sound is small compared to that of light. Therefore, coincidence windows between two spatially separated sensors have to be relatively large.
2. There is substantial acoustic background in the sea (whales!).

These disadvantages limit the applicability of this technique to the very high end of the investigated energy spectrum of the neutrinos. To particle physicists, $>10^{18}$ eV is an extremely high energy. However, we are dealing here with a macroscopic phenomenon, "mini explosions" in which a total energy of $\sim 1$ Joule is released.

In the Mediterranean, the ANTARES telescope infrastructure has been used to study acoustic particle detection [49]. The so-called *AMADEUS* system consists of six "acoustic clusters," each comprising six acoustic sensors (hydrophones) that are arranged at distances of $\approx 1$ m from each other. Figure 1.18a shows one such cluster. The hydrophones use piezo-electric elements for the broad-band recording of signals with frequencies up to 125 kHz (Fig. 1.18b).

At first sight, the South Pole ice cap seems to be a more favorable environment than the Mediterranean Sea for detecting high-energy neutrinos using acoustic signals. The noise level is expected to be much lower, because of the complete absence of sound-emitting life forms. The speed of sound is also much higher in ice than in water, which means that coincidence windows can be much shorter, thus reducing the probability for fake trigger signals. Theoretical estimates put the attenuation length for sound waves with a typical frequency of 20 kHz at 8 km [50], dominated by absorption rather than scattering.

Given these favorable conditions, the option of installing an array of acoustic detectors, to be operated in conjunction with the light detecting IceCube instruments, seemed very attractive, especially because the fiducial volume could be substantially increased, perhaps by several orders of magnitude. For this reason, a subgroup of IceCube researchers set out to measure the relevant parameters in the IceCube environment. This project became known as *SPATS*, the South Pole Acoustic Test Setup.

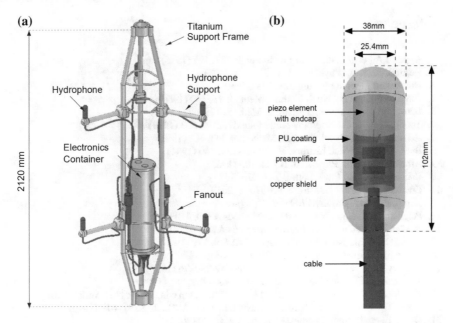

**Fig. 1.18** Drawing of a standard acoustic cluster with hydrophones (**a**), and a schematic drawing of one of the hydrophones (**b**) tested in the ANTARES setup. From: Aguilar, J.A. et al. (2005). *Nucl. Instr. and Meth.* **A626–627**, 128

An array of acoustic transmitters and sensors was set up in some of the IceCube holes, at depths varying from 80 to 500 m below the surface and spaced horizontally up to 690 m [51]. Some results of the tests include [52]:

1. The speed of sound quickly increases with depth and reached a maximum value of ~3,900 m/s at a depth of 200 m.
2. The noise level decreases slightly with increasing depth, it is stable in time and exhibits no correlation with surface conditions (wind, temperature). There is some correlation with human activity, especially if that involves heavy machinery.
3. The relationship between the measured noise floor and the minimum detectable neutrino energy is not clear.
4. The attenuation length was consistently measured, using a variety of methods, sensors, distances and positions, to be ~300 m.

The last result was a great surprise. It is not understood where the calculations went so wrong. But in any case, it has important consequences for the design of a hybrid detector system, which would have to be scaled down considerably in size from the originally envisaged 100 km$^3$ to have any sensitivity at all.

In the meantime, more than 35 years after the idea of acoustic neutrino detection was first proposed, not a single neutrino has ever been observed producing both optical and acoustic signals, neither at the South Pole, nor in the Mediterranean Sea, nor in Lake Baikal (where this technique was also tried).

# References

1. Appel, J.A., et al.: Nucl. Instrum. Methods **127**, 495 (1975)
2. Herb, S.W., et al.: Phys. Rev. Lett. **39**, 252 (1977)
3. Hofstadter, R., et al.: Nature **221**, 228 (1969)
4. Hughes, E.B., et al.: Nucl. Instrum. Methods **75**, 130 (1969)
5. Benvenuti, A., et al.: Nucl. Instrum. Methods **125**, 447 (1975)
6. Young, G.R., et al.: Nucl. Instrum. Methods **A279**, 503 (1989)
7. Acosta, D., et al.: Nucl. Instrum. Methods **A308**, 481 (1991)
8. Beer, A., et al.: Nucl. Instrum. Methods **A224**, 360 (1984)
9. Arnison, G., et al.: Phys. Lett. B **122**, 103 (1983)
10. Banner, M., et al.: Phys. Lett. B **122**, 476 (1983)
11. Ereditato, A., et al.: J. Instrum. **8**, P07002 (2013)
12. Kemp, E.: (2017). arXiv:1709.09385 [hep-ex]
13. Peigneux, J.P., et al.: Nucl. Instrum. Methods **A378**, 410 (1996)
14. Bakken, J.A., et al.: Nucl. Instrum. Methods **A254**, 535 (1987)
15. Ahn, H.S., et al.: Nucl. Instrum. Methods **A410**, 179 (1998)
16. Dong, M.-Y., et al.: Chin. Phys. C **32**, 11 (2008)
17. Akrawy, M.A., et al.: Nucl. Instrum. Methods **A290**, 76 (1990)
18. Anderson, D.F., et al.: Nucl. Instrum. Methods **A290**, 385 (1990)
19. Knoll, G.F.: Radiation Detection and Measurement, 4th edn. Wiley, New York (2010)
20. Doke, T., Masuda, K., Shibamura, E.: Nucl. Instrum. Methods **A291**, 617 (1990)
21. Doke, T., et al.: Nucl. Instrum. Methods **A505**, 199 (2003)
22. Mihara, S.: J. Phys. Conf. Ser. **308**, 012009 (2011)
23. Fukuda, S., et al.: Nucl. Instrum. Methods **A501**, 418 (2003)
24. Halzen, F., Gaisser, ThK: Ann. Rev. Nucl. Part. Sci. **64**, 101 (2014)
25. Ageron, M., et al.: Nucl. Instrum. Methods **A656**, 11 (2011)
26. Collaboration, The Pierre Auger: Nucl. Instrum. Methods **798**, 172 (2015)
27. Aulchenko, V., et al.: J. Phys. Conf. Ser. **587**, 012045 (2015)
28. Michael, D.G., et al.: Nucl. Instrum. Methods **A596**, 190 (2008)
29. Leo, W.R.: Techniques for Nuclear and Particle Physics Experiments. Springer, Berlin (1987)
30. Arnaudon, H., et al.: Nucl. Instrum. Methods **A342**, 558 (1994)
31. Lorenz, E., et al.: Nucl. Instrum. Methods **A344**, 64 (1994)
32. Renker, D., Lorenz, E.: J. Instrum. **4**, P04004 (2009). and references therein
33. Simon, F.: (2018). arXiv:1811.03877 [hep-ex]
34. Wigmans, R.: Calorimetry—Energy Measurement in Particle Physics, 2nd edn. International Series of Monographs on Physics, vol. 168. Oxford University Press, Oxford (2017)
35. Hartjes, F.G., Wigmans, R.: Nucl. Instrum. Meth. **A277**, 379 (1989)
36. Grupen, C., Shwartz, B.: Particle Detectors, 2nd edn. (2008) (Monographs on Particle Physics, Nuclear Physics and Cosmology, vol. 26. Cambridge University Press, Cambridge)
37. Almerio, S., et al.: Nucl. Instr. Meth. **A527**, 329 (2004)
38. Fanti, V., et al.: Nucl. Instrum. Methods **A574**, 433 (2007)
39. Baudis, L.: Ann. Phys. (Berlin) **528**, 74 (2016)
40. Nygren, D.R.: Phys. Scr. **23**, 584 (1981)
41. Sauli, F.: Gaseous Radiation Detectors, Fundamentals and Applications. Monographs on Particle Physics, Nuclear Physics and Cosmology, vol. 36. Cambridge University Press, Cambridge (2014)
42. Agnese, R., et al.: Phys. Rev. D **97**, 022002 (2018)
43. Armengaud, E., et al.: (2017). arXiv:1706.01070 [phys.ins-det]
44. Petricca, F., et al.: (2017). arXiv:1711.07692 [astro-ph]
45. Alduino, C., et al.: Phys. Rev. Lett. **120**, 132501 (2018)
46. Pretzl, K.: Nucl. Instr. Meth. **A454**, 114 (2000)
47. Enss, C. (ed.): Cryogenic Particle Detection. Springer, Berlin (2005)
48. Bevan, S., et al.: Astropart. Phys. **28**, 366 (2007)

49. Aguilar, J.A., et al.: Nucl. Instrum. Methods **A626–627**, 128 (2005)
50. Price, B.J.: J. Geophys. Res. **111**, B02201 (2006)
51. Abbasi, R., et al.: Astropart. Phys. **34**, 382 (2011)
52. Laihem, K.: Nucl. Instrum. Methods **A692**, 192 (2012)

# Part I
# The Basics of Calorimetry

# Chapter 2
# Interactions of Particles with Matter

## 2.1 Introduction

In the absorption process, the energy carried by the incoming particle is eventually distributed in a typically large number of intermediate steps to electrons and (constituents of) nuclei that are part of the absorber structure. Eventually, all the available energy of the incoming particle is shared among a very large number of these so called *shower particles*, each of which carries so little kinetic energy that all that is left for them to do is to ionize or excite the atoms/molecules of the medium they traverse or (in the case of neutrons) scatter off atomic nuclei before being captured by one. The ionization charge and/or the light produced in these processes forms the basis of the calorimeter signals.

In this part, we present the basic facts that determine the performance of calorimeters. In this chapter, the various processes that play a role in the absorption of energetic particles in matter are described. In Chap. 3, the entire absorption process is being considered. And finally, in Chap. 4 some important experimental phenomena that affect the calorimeter signals are discussed.

## 2.2 The Electromagnetic Interaction

The best known energy-loss mechanism contributing to the absorption process is the electromagnetic interaction experienced by charged particles traversing matter. The particles ionize the medium, if their energy is at least sufficient to release the atomic electrons from the Coulomb fields generated by the atomic nuclei. This process also forms the principle on which many particle detectors are based, since the liberated electrons may be collected by means of an electric field and yield an electric signal.

© Springer Nature Switzerland AG 2019
M. Livan and R. Wigmans, *Calorimetry for Collider Physics, an Introduction*,
UNITEXT for Physics, https://doi.org/10.1007/978-3-030-23653-3_2

The em interaction may manifest itself, however, in several other ways:

- Charged particles may excite atoms or molecules without ionizing them. The de-excitation from these metastable states may yield (scintillation) light, which is also fruitfully used as a source of calorimeter signals.
- Charged particles traveling faster than the speed of light characteristic for the traversed medium lose energy by emitting Čerenkov light.
- At high energies, energetic knock-on electrons ($\delta$-rays) are produced.
- At high energies, bremsstrahlung is produced.
- At very high energies, the em interaction may induce nuclear reactions.

## 2.2.1   Electrons and Positrons

Already at energies above 100 MeV, and in many materials even at energies considerably lower than that, by far the principal source of energy loss by *electrons and positrons* is *bremsstrahlung*. In their passage through matter, electrons and positrons radiate photons as a result of the Coulomb interaction with the electric fields generated by the atomic nuclei. The energy spectrum of these photons falls off as $1/E$. It extends, in principle, all the way to the energy of the radiating particle, but in general each emitted photon carries only a small fraction of this energy.

In this process, the electron (or positron) itself undergoes a (usually small) change in direction. This deviation depends on the angle and the energy of the emitted photon, which in turn depend on the strength of the Coulomb field, i.e., on the $Z$ of the absorber material.

These radiative processes, which dominate the absorption of high-energy electrons and positrons, play a role for *any* charged particle traversing matter. However, for heavier charged particles the competition with ionization as the main source of energy loss only starts to play a role at much higher energies. The *critical energy*, $\varepsilon_c$, which may be defined as the energy at which the average energy losses from radiation processes equal those from ionization, is higher by a factor $(m/m_e)^2$, where $m$ and $m_e$ are the particle and the electron mass, respectively. The critical energy of the next-lightest charged particle, the muon ($m_\mu \approx 207 m_e$), is thus about 40,000 times larger than that of the electron. For hadrons such as pions ($m_\pi \approx 273 m_e$), kaons ($m_K \approx 966 m_e$) and protons ($m_p \approx 1836 m_e$), the critical energy is correspondingly larger, but is in practice irrelevant for calorimetry because of the nuclear interactions to which these particles are subject.

The energy loss mechanisms for electrons and positrons are governed by the laws of Quantum Electrodynamics (QED) and can be calculated with a high degree of accuracy. The relative importance of ionization and radiation losses at a given energy depends primarily on the electron density of the medium in which the shower develops. This density is roughly proportional to the (average) $Z$ of the medium, since the number of atoms per unit volume is, within a factor of about two, the same for all materials in the solid state.

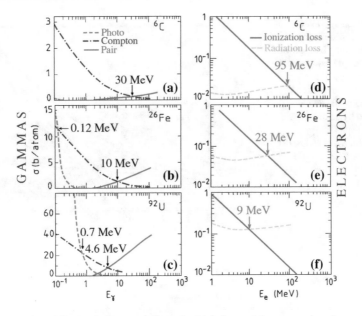

**Fig. 2.1** Cross sections for the processes through which the particles composing electromagnetic showers lose their energy, in various absorber materials. To the left are shown the cross sections for pair production, Compton scattering and photoelectric effect in carbon (**a**), iron (**b**) and uranium (**c**). To the right, the fractional energy losses by radiation and ionization are given as a function of the electron energy in carbon (**d**), iron (**e**) and uranium (**f**) [1]

Results of calculations on the energy loss mechanisms for electrons are shown as a function of energy in Fig. 2.1, for three different absorber materials: carbon ($Z = 6$, Fig. 2.1d), iron ($Z = 26$, Fig. 2.1e) and uranium ($Z = 92$, Fig. 2.1f) [2]. The energy at which the energy losses from ionization equal those from radiation decreases from about 95 MeV for carbon, to 28 MeV for iron, to 9 MeV for uranium.

The Particle Data Group [3] prefers a slightly different definition of the critical energy (at least for electrons), originally formulated by Rossi [4]. In this definition, $\varepsilon_c$ is the energy at which the ionization loss per *radiation length* ($X_0$) equals the electron energy. The radiation length is defined as the distance over which high-energy electrons and positrons lose, on average, 63.2% ($1 - 1/e$) of their energy through radiation. This variable was introduced to describe the development of electromagnetic showers in a material independent way (see Sect. 3.3.1 for details).

$$(\Delta E)_{\text{ion}} = \left[ \frac{dE}{dx} \right]_{\text{ion}} X_0 = E \tag{2.1}$$

This definition would thus be equivalent to the first one if the energy loss to bremsstrahlung were given by

**Fig. 2.2** Energy losses through ionization and bremsstrahlung by electrons in copper. The values for the critical energy following from the two definitions discussed in the text are indicated by arrows. From: Particle Data Group, Tanabashi. M. et al. (2018), *Phys. Rev.* **D98**, 030001

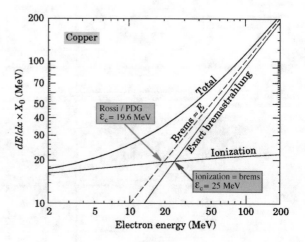

$$\left[\frac{dE}{dx}\right]_{\text{brems}} = \frac{E}{X_0} \tag{2.2}$$

which is true at very high energies, where ionization losses are negligible, but which is only an approximation in the energy regime near $\varepsilon_c$. The difference between the two definitions is illustrated in Fig. 2.2, for electrons in copper. In this figure, Eq. 2.2 is represented by the dashed line. Using this alternative definition, the PDG has fitted the $dE/dx$ data tabulated by Pages [2] and gives the following expressions for the critical energy:

$$\varepsilon_c = \frac{610 \text{ MeV}}{Z + 1.24} \tag{2.3}$$

for materials in the solid or liquid phase, and

$$\varepsilon_c = \frac{710 \text{ MeV}}{Z + 0.92} \tag{2.4}$$

for gases. These formulae fit the data from the mentioned $dE/dx$ tables to within ~4%, with the largest deviations occurring at the highest $Z$ values. For example, for uranium, Eq. 2.3 gives $\varepsilon_c = 6.54$ MeV, while the data tabulated in [2] fulfill Rossi's condition at an energy of 6.75 MeV.

The $\varepsilon_c$ values found in this way are systematically smaller than the ones following from the other definition, where $\varepsilon_c$ is the energy at which ionization losses equal radiation losses. The differences range from ~15% for carbon to ~35% for uranium (see also Fig. 2.2).

In the following, we will use the definition of Rossi and the PDG.

## 2.2.2 Muons Traversing Dense Material

The other particles subjected to only the em interaction, the muons, behave, at the same energies, in a very different way than electrons and positrons. Whereas radiative processes dominate the energy loss of the latter particles, muons traversing dense material lose their energy up to very high energies (100 GeV or higher) primarily through ionization and $\delta$-rays. These mechanisms account for energy losses of typically only 1–2 MeV g$^{-1}$cm$^2$ and, therefore, it takes very substantial amounts of material to absorb high-energy muons.

For this reason, experiments in which cosmic muons constitute a major source of undesirable background have to be located in deep mines or under high mountains, since these muons may sometimes penetrate several kilometers of the Earth's crust. For the same reason, the CERN high-energy neutrino beam that was used for many experiments in the West Area (1963–1998) was equipped with a 300 m long iron shield. The neutrinos were produced from pion and kaon decay ($\pi$, $K \rightarrow \nu_\mu \mu$) and the muons had to be absorbed in the space between the production target and the neutrino detectors. In iron, muons lose energy at a rate of about 1.1 GeV/m (about a factor of three higher than in the soil of which the CERN site is composed).

Higher-order QED processes, such as bremsstrahlung and direct $e^+e^-$ pair production, do also occur in muon absorption. However, compared with electrons, they are suppressed by a scale factor of $(m_\mu/m_e)^2 \approx 40,000$. Therefore, the critical energy at which muons lose, on average, equal amounts of energy through radiation and ionization is at least 200 GeV. Just as for electrons, the contribution of these higher-order QED processes to a muon's energy loss is strongly $Z$ dependent [5]. For example, the average energy loss of 500 GeV muons in lead increases (with respect to ionization losses) by a factor of 5.8 because of these effects. In iron, this factor amounts to 2.5 and in aluminium 1.8.

At energies below 100 GeV, the average energy loss is primarily determined by ionization, in all absorber materials. The mean energy loss per unit path length, $\langle dE/dx \rangle$, is given by the Bethe–Bloch formula [4]:

$$- \langle dE/dx \rangle = K z^2 \frac{Z}{A} \frac{1}{\beta^2} \left[ \frac{1}{2} \ln \frac{2 m_e c^2 \beta^2 \gamma^2 T_{\max}}{I^2} - \beta^2 - \frac{\delta}{2} \right] \quad (2.5)$$

in which $T_{\max}$ represents the maximum kinetic energy that can be imparted to an electron in a single collision, $I$ is the mean excitation energy of the absorber material, $\delta$ a correction term describing the *density effect*, and the proportionality constant $K$ equals $4\pi N_A r_e^2 m_e c^2$.

The quantity $\langle dE/dx \rangle$, which is often referred to as the *specific ionization* or the *ionization density*, has a characteristic energy dependence, which is governed by the product of the velocity ($\beta$) and the Lorentz factor ($\gamma$) of the particles (Fig. 2.3). For relativistic muons, $\langle dE/dx \rangle$ falls rapidly with increasing $\beta$, reaches a minimum value near $\beta = 0.96$, and then exhibits what is called the relativistic rise, to level off at values of 1–2 MeV g$^{-1}$ cm$^2$ in most materials. Muons, or other particles with unity

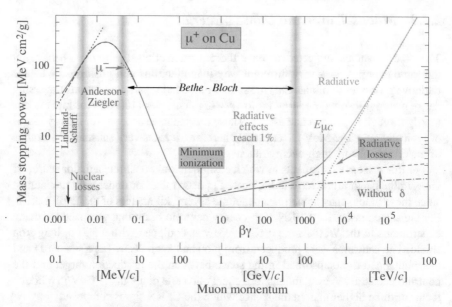

**Fig. 2.3** The average energy loss per unit path length $(-\langle dE/dx \rangle)$ for positive muons in copper, given as a function of the product of the Lorentz variables $\beta\gamma$. For muon momenta in the range from $\sim$5 MeV/$c$ to $\sim$50 GeV/$c$, this energy loss is well described by the Bethe-Bloch formula (Eq. 2.5.) From: Particle Data Group, Tanabashi. M. et al. (2018), *Phys. Rev.* **D98**, 030001

charge such as pions, with an energy corresponding to that at which the $\langle dE/dx \rangle$ curve reaches its minimum, are called *minimum ionizing particles*, or *mips*,

In relatively thin amounts of material, such as those represented by a typical calorimeter, the total energy loss $\Delta E/\Delta x$ may differ substantially from the value calculated on the basis of $\langle dE/dx \rangle$. This is because of the relatively small number of collisions with atomic electrons, and the very large fluctuations in energy transfer that may occur in such collisions. Therefore, the energy loss distributions measured with (thin) calorimeters reach their maximum (i.e., most probable) value in general below the value calculated on the basis of $\langle dE/dx \rangle$ and have a long tail toward large energy losses, the so-called Landau tail [6]. Only for very substantial amounts of matter, e.g., equivalent to 100 m of water, the energy loss distribution becomes approximately Gaussian.

It should be emphasized that the Bethe-Bloch formula does not only apply to muons, but to the ionization losses of all charged particles. In that sense, it is an extremely important formula for calorimetry in general, since the signals produced by calorimeters are the result of processes in which the atoms of the absorber material are excited. The $\beta^{-2}$ dependence of the energy loss for non-relativistic particles turns out to have crucial consequences for the signals from hadronic showers.

### 2.2.3 Photon Interactions

The quantum of the em interaction, the photon ($\gamma$) is mainly affected by four different processes: the photoelectric effect, coherent (Rayleigh) scattering, incoherent (Compton) scattering and electron–positron pair production.

#### 2.2.3.1 Photoelectric Effect

At low energies, this is the most likely process to occur. In this process, an atom absorbs the photon and emits an electron. The atom, which is left in an excited state, returns to the ground state by the emission of Auger electrons or X-rays. The photoelectric cross section is extremely dependent on the available number of electrons, and thus on the $Z$ value of the absorber material. It scales with $Z^n$, with the power $n$ between 4 and 5. The photoelectric cross section varies with the photon energy as $E^{-3}$, so that this process rapidly loses its importance as the energy increases. In uranium, the highest-$Z$ material that can be used for calorimeter construction, the cross section for photoelectric effect is dominating for energies below 700 keV, for iron inelastic scattering already starts to dominate above 100 keV (see Fig. 2.5).

#### 2.2.3.2 Rayleigh Scattering

This (coherent) process is also important at low energies. In this process, the photon is deflected by the atomic electrons. However, the photon does *not* lose energy. Therefore, Rayleigh scattering affects the spatial distribution of the energy deposition, but it does not contribute to the energy deposition process itself.

#### 2.2.3.3 Compton Scattering

In the Compton process, a photon is scattered by an atomic electron, with a transfer of momentum and energy to the struck electron that is sufficient to put this electron in an unbound state.

Figure 2.4 illustrates this scattering process. Applying the laws of energy and momentum conservation, the relations between the different kinematic variables (energy transfer, scattering angles) can be derived in a straightforward manner. For example, when $\zeta$ is defined as the photon energy in units of the electron restmass ($\zeta = E_\gamma / m_e c^2$), the scattering angles of the electron ($\phi$) and the photon ($\theta$) are related as

$$\cot \phi = (1 + \zeta) \tan \frac{\theta}{2} \tag{2.6}$$

In all but the highest-$Z$ absorber materials, Compton scattering is by far the most likely process to occur for $\gamma$s in the energy range between a few hundred keV and

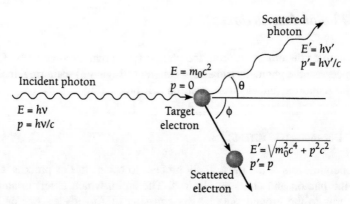

**Fig. 2.4** The Compton scattering process [1]

~5 MeV (see Fig. 2.5). As we shall see in Sect. 3.4, typically at least half of the total energy is deposited by such $\gamma$s in the absorption process of multi-GeV electrons, positrons or photons. Compton scattering is therefore a very important process for understanding the fine details of calorimetry.

The angular distribution of the Compton recoil electrons exhibits a preference for the direction of the incoming photons ($\cos \phi = 1$), but there is also a substantial isotropic component in the forward hemisphere (the requirements of momentum and energy conservation prevent the electrons from being scattered in the backward hemisphere).

Since the photoelectric effect, in which the photon is absorbed and thus disappears, only plays a role at low energies, many $\gamma$s in the MeV energy range are absorbed in a *sequence* of Compton scattering processes, in which the photon energy is reduced in a number of steps down to the point where the final absorption in a photoelectric process occurs. In each step, an amount of energy equal to

$$T = E_\gamma \frac{\zeta(1 - \cos\theta)}{1 + \zeta(1 - \cos\theta)} \tag{2.7}$$

is transferred to the struck electron. In this process, the angular preference still visible for the *first* scattering in this sequence quickly disappears. Most of the Compton- and photoelectrons produced in this sequential absorption process are isotropically distributed with respect to the direction of the initial $\gamma$.

The cross section for Compton scattering is much less dependent on the Z value of the absorber material than the cross section for photoelectric effect. The Compton cross section is almost proportional to Z, i.e., proportional to the number of target electrons in the nuclei.

As for the photoelectric effect, the cross section for Compton scattering decreases with increasing photon energy, albeit much less steeply: $\sigma \sim 1/E$. Therefore, above a certain threshold energy, Compton scattering becomes more likely than photoelectric

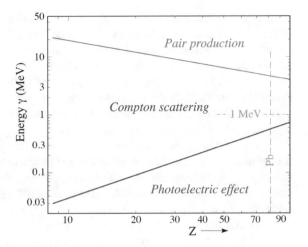

**Fig. 2.5** The energy domains in which photoelectric effect, Compton scattering and pair production are the most likely processes to occur, as a function of the $Z$ value of the absorber material [1]

absorption. This threshold ranges from 20 keV for carbon ($Z = 6$) to 700 keV for uranium ($Z = 92$). The values for other elements can be derived from Fig. 2.5.

### 2.2.3.4 Pair Production

At energies larger than twice the electron rest mass, a photon may create, in the field of a charged particle, an electron–positron pair. These particles produce bremsstrahlung radiation as well as ionization along their paths. The electron is eventually absorbed by an ion, while the positron annihilates with an electron. In the latter process, two new photons are produced, each with an energy of 511 keV, the electron restmass, if the annihilation takes place when the positron has come to rest.

Typically, more than 99% of the $\gamma \rightarrow e^+e^-$ conversions are caused by *nuclear* electromagnetic fields. For low-$Z$ elements and at high energies, $e^+e^-$ creation in the fields of the atomic electrons also contributes significantly to the total pair production cross section.

The cross section for pair production rises with energy and reaches an asymptotic value at very high energies ($>1$ GeV). This cross section is related to the *radiation length* of the absorber material (see Sect. 3.3.1).

The relative importance of the processes through which photons are absorbed depends strongly on the photon energy and on the electron density ($\sim Z$) of the medium. This is illustrated in Fig. 2.1, which shows the cross sections for these three processes as a function of energy in carbon ($Z = 6$, Fig. 2.1a), iron ($Z = 26$, Fig. 2.1b) and uranium ($Z = 92$, Fig. 2.1c).

Since the cross sections for the photoelectric effect and for Compton scattering decrease with energy, and the cross section for pair production increases, pair production is the most likely process to occur at high energies.

The photoelectric effect dominates at low energies. Compton scattering is the process of choice in some intermediate regime (Fig. 2.5). The higher the $Z$ value of the absorber, the more limited the role of Compton scattering in the em absorption process. In uranium, Compton scattering dominates for energies between 0.7 MeV and 4.6 MeV. For iron, this interval is extended to 0.12 MeV on the low-energy side and to 10 MeV on the high-energy side. For low-$Z$ materials such as carbon, pair production only takes over above 30 MeV, while the photoelectric effect is only significant at energies below 50 keV.

Figure 2.1 also shows that the *total cross section* for photon interactions exhibits a minimum value near the energy where the probabilities for Compton scattering and pair production are about equal. For high-$Z$ materials, which have the highest total cross sections and are therefore best suited for shielding against $\gamma$-rays, this minimum occurs for energies around 3 MeV. It may seem quite amazing that it takes less material to shield effectively against $\gamma$s of 10–20 MeV than against 3 MeV ones. However, this peculiarity is by no means unique to the MeV range of the em spectrum. For example, the Earth's atmosphere is transparent to visible light, but light with a shorter wavelength (higher energy, e.g., ultraviolet light or X-rays) is effectively absorbed by it.

There are similarities and differences between the ways photons and charged particles interact with matter and are absorbed as a result of these interactions. In both cases, inelastic interactions with the atomic electrons play an important role. However, the cross sections for these processes are very different. For charged particles, the cross sections are typically $10^7 - 10^8$ barns, while the cross sections for photon interactions are 3–5 orders of magnitude smaller, varying from $10^4$ b for photoelectric effect at low energies in high-$Z$ absorber materials to less than $10^2$ b for pair production.

As a result of these differences, the absorption of high-energy $\gamma$s appears to proceed in a fundamentally different way than for electrons and positrons. When these charged particles traverse matter, they lose energy in a continuous stream of events in which atoms or molecules are ionized and photons are radiated away. A multi-GeV electron traversing one cm of lead typically radiates thousands of photons. Some of these photons may have energies in excess of 1 GeV, but the overwhelming majority of these photons are very soft, with energies in the eV–keV–MeV range.

On the other hand, a multi-GeV photon may penetrate the same thickness of lead *without being affected at all*. For such high-energy photon interactions, we may apply the concept of the *mean free path*, 7.2 mm in the case of lead. Therefore, the probability that the mentioned photon does interact (i.e., convert into an $e^+e^-$ pair) in one cm of lead is $[1 - \exp(-10/7.2)]$, or about 75%. For electrons, this concept has no meaning. What we can say is that the original electron has lost, on average, about 83% of its energy after traversing this material.

### 2.2.3.5 Photonuclear Reactions

At energies in the range 5–20 MeV, a modest role may be played by photonuclear reactions, e.g., $\gamma n$, $\gamma p$ or photo-induced nuclear fission. The cross sections for such reactions reach a maximum value at the so-called giant dipole resonance, when the photon energy is approximately equal to the *marginal binding energy* of the proton or neutron, i.e., the difference in nuclear binding energy between the target nucleus and the nuclei with one nucleon less. The cross sections for these processes usually do not exceed 1% of the total cross section for the processes mentioned in the previous subsections [7].

## 2.3 The Strong Interaction

So far, we have only considered electromagnetic interactions between shower particles and the (em fields in the) absorbing medium. We now turn our attention to the absorption of high-energy hadrons, in which the strong interactions between the shower particles and the nuclei of the absorbing medium also play an important role.

Because of the nature of the strong interaction, hadronic showers are much more complicated than electromagnetic ones. The variety of processes that may occur, both those at the particle level and those involving the struck nucleus, is much larger.

When discussing the particles that play a role in em showers (Sect. 2.2), we saw an important difference between the absorption of photons and electrons. Electrons lose their energy in a continuous stream of events, in which atoms of the traversed medium are ionized and bremsstrahlung photons are emitted. On the other hand, photons may penetrate a considerable amount of matter without losing any energy, and then interact in a manner that may change their identity (i.e., the photon may turn into a $e^+e^-$ pair).

When a high-energy hadron penetrates a block of matter, some combination of these phenomena may occur. When the hadron is charged, it will ionize the atoms of the traversed medium, in a continuous stream of events, in much the same way as a muon of the same energy would do (Sect. 2.2.2). However, in general, at some depth, the hadron encounters an atomic nucleus with which it interacts strongly.

In the nuclear interaction, the hadron may change its identity dramatically. It may, for example, turn into fifteen new hadrons. Also the struck nucleus changes usually quite a bit in such a reaction. It may, for example, lose ten neutrons and three protons in the process and end up in a highly excited state, from which it decays by emitting several $\gamma$-rays.

Neutral hadrons do not ionize the traversed medium. For these particles, nuclear reactions are *the only* option for losing energy. This is in particular true for neutrons, which are abundantly produced in hadronic shower development. As a result, neutrons deposit their kinetic energy in ways very different from those for the charged shower particles, with potentially very important implications for calorimetry.

**Table 2.1** The specific ionization energy loss of minimum ionizing particles in various absorber materials, and the average energy lost by minimum ionizing protons over a distance of one nuclear interaction length. Data from [3]

| Absorber | Z | $dE/dx$ (mip) (MeV g$^{-1}$cm$^2$) | $\lambda_{int}$ (g cm$^{-2}$) | $\Delta E/\lambda_{int}$ (MeV) |
|----------|-----|------|-------|-----|
| Carbon | 6 | 1.742 | 85.8 | 149 |
| Aluminum | 13 | 1.615 | 107.2 | 173 |
| Iron | 26 | 1.451 | 132.1 | 192 |
| Copper | 29 | 1.403 | 137.3 | 193 |
| Tin | 50 | 1.263 | 166.7 | 211 |
| Tungsten | 74 | 1.145 | 191.9 | 220 |
| Lead | 82 | 1.122 | 199.6 | 224 |
| Uranium | 92 | 1.081 | 209.0 | 226 |

## 2.3.1 Particle Production in the Absorption of High-Energy Hadrons

Some of the particles produced in the absorption of energetic hadrons, in particular $\pi^0$s and $\eta$s, decay through the electromagnetic interaction: $\pi^0, \eta \to \gamma\gamma$. Therefore, hadron showers generally contain a component that propagates purely electromagnetically. The fraction of the initial hadron energy converted into $\pi^0$s and $\eta$s (which we will call $f_{em}$ in the following) varies strongly from event to event.

On average, approximately one-third of the mesons produced in the interaction are $\pi^0$s. The *charged* hadrons produced in the collision may induce a nuclear interaction themselves after having traveled, on average, an interaction length in the material. The average energy a minimum ionizing particle loses per unit $\lambda_{int}$ through ionization in this process is listed for various absorber media in Table 2.1. It amounts to ~150 MeV for low-$Z$ materials such as carbon, and increases with $Z$ to reach about 225 MeV for high-$Z$ absorber materials, such as lead or uranium. It should be remarked that the values listed in Table 2.1 are for protons. Also, the values in Table 2.1 are for *minimum ionizing particles*. In reality, the specific ionization loss of the pions produced in the shower development is typically somewhat larger than the mip value. As a result of these effects, pions produced in hadronic showers developing in uranium lose, on average, some 300 MeV by ionization, before inducing a nuclear reaction. For iron and aluminium, the losses are 15–25% smaller. These differences constitute a significant contribution to the $Z$ dependence of the hadronic absorption process.

## 2.3.2 Nuclear Reactions in the Absorption of High-Energy Hadrons

### 2.3.2.1 Nuclear Spallation Reactions

When an incoming high-energy hadron strikes an atomic nucleus, the most likely process to occur is *spallation*. Spallation is usually described as a two-stage process: a fast intranuclear cascade, followed by a slower evaporation stage. The incoming hadron makes quasi-free collisions with nucleons inside the struck nucleus. The affected nucleons start traveling themselves through the nucleus and collide with other nucleons. In this way, a cascade of fast nucleons develops. At this stage, pions and other unstable hadrons may also be created if the transferred energy is sufficiently high. Some of the particles taking part in this cascade reach the nuclear boundary and escape. Others get caught and distribute their kinetic energy among the remaining nucleons in the nucleus.

The second step of the spallation reaction consists of a de-excitation of the intermediate nucleus. This is achieved by evaporating a certain number of particles, predominantly free nucleons, but sometimes also $\alpha$s or even heavier nucleon aggregates, until the excitation energy is less than the binding energy of one nucleon. The remaining energy, typically a few MeV, is released in the form of $\gamma$-rays. In very heavy nuclei, e.g., uranium, the intermediate nucleus may also fission.

Much experimental information on spallation reactions has been accumulated during the past 70 years. Rudstam [8] has given a useful empirical formula, valid within broad limits either of energies ($>50$ MeV) or of atomic mass ($A > 20$), which gives a satisfactory description of spallation cross sections. When a particle of energy $E$ hits a target with atomic mass $A_T$, the relative cross sections $\sigma$ for the production of spallation products ($Z_f, A_f$) are given by the relation

$$\sigma(Z_f, A_f) \sim \exp\left[-P(A_T - A_f)\right] \times \exp\left[-R|Z_f - SA_f + TA_f^2|^{3/2}\right] \quad (2.8)$$

in which $E$ is expressed in MeV and the parameters $P$, $R$, $S$ and $T$ have the following values: $P = 20E^{-0.77}$ for $E < 2100$ MeV, $P = 0.056$ for $E > 2100$ MeV, $R = 11.8A_f^{-0.45}$, $S = 0.486$, $T = 0.00038$.

Figure 2.6 shows the cross sections for nuclides that can be produced from $^{238}$U spallation induced by a 2 GeV hadron, computed with this formula. Hundreds of different reactions occur with comparable probability. The largest cross section for an exclusive reaction amounts to only $\sim 2\%$ of the total spallation cross section, and there are about 300 different reactions that contribute more than 0.1% to the total spallation cross section! This example illustrates the enormous diversity of processes that may occur in the nuclear sector of the hadronic interactions.

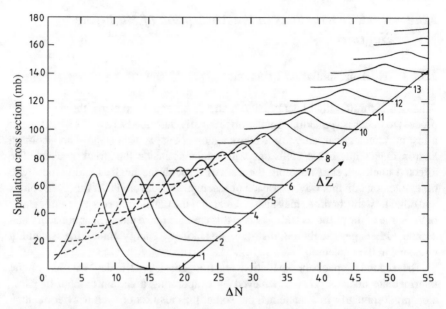

**Fig. 2.6** Cross sections for nuclides produced by spallation of $^{238}$U, induced by a 2 GeV hadron. The final-state nuclide is defined by the number of protons ($\Delta Z$) and neutrons ($\Delta N$) released from the target nucleus [1]

### 2.3.2.2 Nuclear Binding Energy

In these spallation reactions, considerable numbers of nucleons may be released from the nuclei in which they were bound. This was beautifully illustrated by emulsion and bubble-chamber pictures from the early days of high-energy physics, which frequently showed "nuclear stars," nuclei that literally exploded in the collisions.

An example of such a process is depicted in Fig. 2.7, which shows (the results of) a collision between a 30 GeV/$c$ proton and an atomic nucleus in a photographic emulsion. About 20 densely ionizing particles, presumably (almost) all protons, are produced in this reaction. The picture shows that these particles are more or less isotropically emitted from the struck nucleus. Several other, much less dense ionization tracks emerge from the collision to the left, i.e., roughly following the direction of the incoming projectile, which enters the picture from the right-hand side. Most likely, these tracks represent pions and fast spallation protons. Not visible in this picture are the neutrons, considerable numbers of which undoubtedly were also released in this interaction.

The energy needed to release these nucleons, i.e., the nuclear binding energy, is lost for calorimetric purposes, it does not contribute to the calorimeter signal, it is *invisible*. There is a large variety of processes that may occur in hadronic shower development and event-to-event fluctuations in the invisible energy fraction are very large. In one extreme case, an incoming $\pi^+$ may strike a neutron and cause the

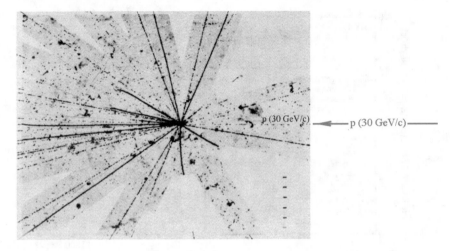

**Fig. 2.7** A proton-nucleus interaction in a nuclear emulsion stack. Photograph courtesy CERN

following reaction: $\pi^+ n \to \pi^0 p$, transferring almost all its kinetic energy to the $\pi^0$ (charge exchange reaction). In that case, the invisible energy is almost zero, since the $\pi^0$ decays in two $\gamma$s which decay electromagnetically and the proton loses its (small) energy through ionization of the medium in which the reaction takes place, leaving only the nuclear binding energy of one escaping nucleon unaccounted for. In other extreme cases, invisible energy may consume some 60% of the total available energy. As is shown later on in this subsection, invisible energy accounts, on average, for 30–40% of the non-em shower energy, i.e., energy that is not carried by $\pi^0$s or other electromagnetically decaying particles produced in the shower development.

The large event-to-event fluctuations in visible energy have obviously direct consequences for the precision with which hadronic energy can be measured in calorimeters. Because of these fluctuations, which have no equivalent in electromagnetic shower development processes, the energy resolution of hadron calorimeters is usually considerably worse than the em energy resolution. This problem, and possible solutions, are discussed in detail in Chaps. 7 and 8.

### 2.3.2.3 Spallation Nucleons

In the spallation reactions with absorber nuclei that take place in the absorption of energetic hadrons, large numbers of nucleons and nucleon aggregates such as $\alpha$ particles are produced. Equation 2.8 provides quantitative information about this process. For example, in the reactions of 1 GeV hadrons and $^{208}_{82}$Pb nuclei, on average, 2.7 protons and 12.8 neutrons are produced. The large discrepancy between the numbers of protons and neutrons released in these reactions is even more striking at lower incident energies. Figure 2.8a shows the average numbers of protons and

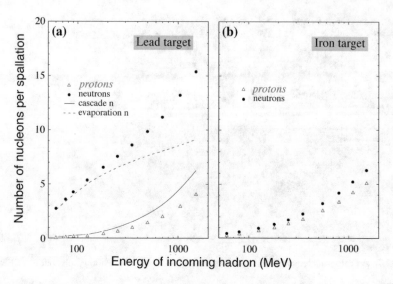

**Fig. 2.8** The average numbers of protons and neutrons produced in spallation reactions on $^{208}_{82}$Pb (*a*) or $^{56}_{26}$Fe (*b*), as a function of the energy of the incoming hadron. The neutrons are split up in an evaporation and a cascade component [1]

neutrons as a function of the incident kinetic energy $E$. For energies smaller than 200 MeV, the probability that at least one proton is emitted drops below 50%. However, on average, 7 neutrons still come off at 200 MeV.

This indicates that the protons that are produced in the spallation processes on lead are almost exclusively produced in the fast cascade step. In the evaporation stage of the reaction, almost all emitted nucleons are neutrons. This is not surprising, since the Coulomb barrier for protons in a lead nucleus is $\sim$12 MeV. Therefore, in the evaporation stage, where fragments are released with a kinetic energy of typically a few MeV (some fraction of the binding energy per nucleon, which amounts to $\sim$7.9 MeV in lead), not many charged particles are expected to emerge from the nucleus. In the fast cascade step, protons and neutrons are emitted in a ratio that, on average, reflects the numerical presence of these nucleons in the target nuclei. In $^{208}_{82}$Pb, one may thus expect for every cascade proton about 1.5 cascade neutrons (126/82).

These considerations make it possible to split the total nucleon production in the spallation reactions induced by our 1 GeV hadrons ($\sim$16 nucleons) into a cascade component and an evaporation component: 9 evaporation neutrons and 7 cascade nucleons (2.8 protons and 4.2 neutrons). The cascade nucleons, in particular the cascade neutrons, are likely to induce themselves new spallation reactions, further increasing the numbers of evaporation neutrons.

The distributions of the numbers of protons and neutrons released from their nuclear environment in spallation reactions are markedly different when the lead target is replaced by an iron one. The dependence of the *average* numbers of such protons and neutrons on the energy of the incident hadron is given in Fig. 2.8b. Among the most characteristic differences, we mention:

- The strong *asymmetry* between protons and neutrons found in the case of lead is almost absent for reactions with iron nuclei.
- The total *number* of nucleons released in collisions with iron nuclei is considerably smaller than that for collisions with lead nuclei at the same energy.

The proton/neutron asymmetry in lead is a consequence of the Coulomb barrier ($\sim$12 MeV), which prevents protons from being emitted by an excited nucleus in the evaporation stage. In iron, this barrier is considerably lower ($\sim$5 MeV). Therefore, the probabilities for an excited Fe nucleus to emit a proton or a neutron are not very different from each other.

The particles emitted in the evaporation stage of the reactions are emitted isotropically, but the cascade particles have a dominating momentum component along the direction of the incoming particle. Therefore, the residual target nucleus undergoes a net recoil, in which it acquires a kinetic energy of the order of $m/M$, where $m$ and $M$ denote the total mass of the cascade nucleons and the residual nucleus, respectively. This recoil energy will, in general, *not* result in a measurable calorimeter signal and therefore has to be considered part of the invisible component of the shower energy.

Let us now return to the earlier example of 1 GeV hadrons striking $^{208}_{82}$Pb nuclei and examine what happens to the energy in the "average" spallation reaction. In order to release the 16 nucleons from the lead nucleus, $16 \times 7.9 = 126$ MeV of nuclear binding energy has to be provided. The 9 evaporation neutrons carry a total of $\sim$27 MeV of kinetic energy (for $T = 2$ MeV, see Fig. 2.9). The remaining $1000-153 = 847$ MeV is shared among the target nucleus (30 MeV recoil energy) and 7 cascade nucleons, about 117 MeV each. The range of 117 MeV protons in lead amounts to $\sim$2 cm [9], considerably less than the nuclear interaction length (17 cm). Therefore, these cascade protons are most likely to lose their energy by ionizing lead atoms.

The (on average, 4.2) cascade neutrons, on the other hand, will induce new spallation reactions. Especially in high-$Z$ materials such as lead, they typically initiate nuclear reactions of the type $^{A}_{Z}X$ $(n, yn)$ $^{A'}_{Z}X$, with $A' = A - y + 1$, in which the total number of evaporation neutrons produced in the shower absorption process thus increases by $(y - 1)$. In practice, almost all neutrons that are present in the absorber structure a few nanoseconds after the start of the shower process are thus of the evaporation type.

### 2.3.2.4 Evaporation Neutrons

A significant fraction of the hadronic shower energy is thus carried by large numbers of soft neutrons. The absorption of these neutrons in dense material proceeds very differently from that of the other types of shower particles encountered so far. Electrons, photons, charged mesons and protons are all subject to the electromagnetic interaction. Neutrons depend entirely on the strong (and sometimes the weak) interaction in order to be absorbed in matter. This has very important consequences for

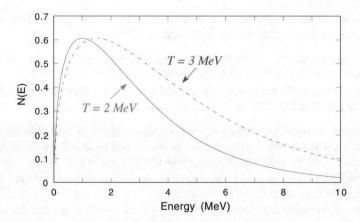

**Fig. 2.9** Kinetic energy spectrum of evaporation neutrons, produced according to a Maxwell distribution with a temperature of 2 MeV. For comparison, the spectrum for a temperature of 3 MeV is given as well [1]

calorimetry. These consequences may range from very beneficial to very detrimental [10] and are extensively discussed in Chaps. 3 and 7. In this chapter we concentrate on the neutron spectra and on the absorption mechanisms.

The kinetic energy spectrum of the evaporation neutrons is usually described by a Boltzmann–Maxwell distribution

$$\frac{dN}{dE} = \sqrt{E} \exp(-E/T) \tag{2.9}$$

with a temperature $T$ of about 2 MeV, so that the average kinetic energy of these neutrons amounts to about 3 MeV at production (see Fig. 2.9).

This means that these neutrons have obtained a kinetic energy that is typically of the order of one-third to one-half of the binding energy that confined them to their parent nuclei before the shower development occurred.

Experimental measurements have revealed that the numbers of neutrons produced in the hadronic absorption process are large. Leroy et al. [11] measured the production rates of *thermalized* neutrons in high-energy hadron showers, by analyzing the induced radioactivity resulting from neutron-capture reactions in the absorber material. They found rates of ∼20 neutrons per GeV of energy in lead and up to 60 neutrons per GeV in $^{238}$U, where nuclear fission causes a significant multiplication of the neutron production rates.

Based on knowledge of the spectrum of the neutrons, the average total kinetic energy carried by these neutrons, and the fluctuations in this total kinetic energy may be calculated. Some results of these calculations, for the energy spectrum given in Fig. 2.10, are shown in Fig. 2.10.

Because of the Central Limit Theorem and because of the large numbers of neutrons involved, the fluctuations in this total kinetic energy are relatively small. For

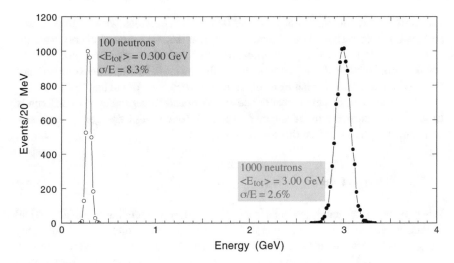

**Fig. 2.10** Distribution of the total kinetic energy carried by 100 and 1,000 evaporation neutrons [1]

example, one thousand neutrons, a number typically produced in the absorption of a 50 GeV hadron in lead, carry a total kinetic energy of 3 GeV ±78 MeV.

Not surprisingly, it turns out that there is a clear correlation between this total kinetic energy carried by the neutrons produced in the shower development and the total amount of *invisible* energy, which is primarily determined by the *number* of target nucleons released in the development of the hadron shower. In Chap. 8, it is shown how this correlation can be exploited to improve the calorimetric performance.

### 2.3.3 The Interactions of Neutrons with Matter

In this subsection we discuss the various mechanisms through which the evaporation neutrons lose their kinetic energy and are eventually absorbed in dense matter.

#### 2.3.3.1 Elastic Neutron Scattering

At energies between a few eV and approximately 1 MeV, elastic scattering is by far the dominant, if not the only, energy loss mechanism for neutrons. The cross sections for elastic neutron scattering in this energy range are large, usually several barns, which implies mean free paths between collisions of typically a few cm.

The energy fraction $f$ lost by neutrons in collisions with nuclei with atomic number $A$ (approximately proportional to the mass) varies between 0 (for glancing collisions) and $4A/(A + 1)^2$, the kinematic limit for central collisions. The average values of $f$ amount thus to 50, 3.4 and 0.96% for collisions with hydrogen, iron and lead, respectively. It is of course no surprise that, in terms of energy loss, the elastic scattering process is most efficient in hydrogen. A large cross section and a considerable fraction of energy lost in each collision make hydrogen-rich compounds the material of choice for neutron shielding purposes, e.g., in nuclear reactors.

As a result, neutrons in the mentioned energy bracket are sampled very differently from charged particles in calorimeter structures containing hydrogen in the active components. This is further discussed in Sect. 8.2.

### 2.3.3.2  Neutron Capture

When the neutrons generated in hadronic shower development have lost (almost) all of their kinetic energy in collisions with the target material, one of two things may happen: they decay, or they get captured by an atomic nucleus. Since the timescale for the first process is very long (average lifetime $\sim$15 min) and the cross section for the second process usually large, capture is much more likely to occur. When a neutron is captured by an atomic nucleus, the nuclear binding energy that had to be supplied to the nucleus when the neutron was released (invisible energy) is gained back. The excited "compound" nucleus usually gets rid of this excess energy by emitting $\gamma$-rays. In some light nuclei, such as $^6$Li and $^{10}$B, the capture of a neutron may be followed by the emission of an $\alpha$ particle.

The neutron capture process is distinctly different from the processes through which charged particles, such as electrons and protons, get absorbed in the shower development process. After losing their kinetic energy through ionization of the calorimeter materials, these charged particles just become part of the absorbing structure, while the neutrons transform an absorber nucleus into another type of nucleus.

### 2.3.3.3  The Production of $\alpha$ Particles

Another process which illustrates that neutrons may be sampled very differently from charged particles takes place at energies between 3 and 20 MeV. At these energies, neutrons frequently release $\alpha$ particles from the nuclei with which they interact, for example through $(n, \alpha)$ reactions. This is particularly true for $^{12}$C, a key ingredient of organic materials such as plastics or gases used in wire chambers. Neutrons above 10 MeV may split this nucleus into three $\alpha$ particles, the reverse process of the one that starts the CNO cycle in aging stars. This process alone accounts for $\sim$60% of the inelastic $n^{12}$C cross section. In other light gases such as oxygen and fluorine, $\alpha$ production is also quite abundant, while in higher-$Z$ materials like iron or copper, $\alpha$ production takes place in only $\sim$3% of the inelastic reactions.

When produced in wire chambers, such $\alpha$s may give rise to signals that are orders of magnitude larger than the ones caused by minimum ionizing particles and, therefore, it may be an important process in calorimeters that use such wire chambers for generating the signals (Sect. 8.2).

#### 2.3.3.4 Inelastic Neutron Scattering

The fact that the energy loss process of neutrons in a given material is extremely dependent on subtleties of that material's nuclear structure, becomes also clear when we examine the role of inelastic scattering.

In this process, part of the neutron's kinetic energy is used to bring a nucleus in an excited state. The nucleus releases this excitation energy in the form of one or several $\gamma$s, whose (combined) energy corresponds to the energy loss of the neutron. The contribution of this process to the energy loss of the neutrons produced in the calorimeter depends completely on details of the nuclear level structure.

In some materials, e.g., lead, it becomes insignificant below energies as high as 2.6 MeV (because it takes that much energy to bring the most abundant lead isotope, $^{208}$Pb, from its ground state into the lowest excited state), in other materials it continues to play a role down to energies well below 1 MeV. For example, the first excited state of the most abundant isotope of iron, $^{56}$Fe, is located 0.85 MeV above the ground state. The cross section for inelastic scattering processes of the type $(n, n'\gamma)$, in which neutrons in the energy range of 1–6 MeV lose 0.85 MeV, is more than one barn [12]. This is one of the reasons why steel-reinforced concrete is a good shielding material for MeV-type neutrons.

# References

1. Wigmans, R.: Calorimetry—Energy Measurement in Particle Physics, 2nd edn. International Series of Monographs on Physics, vol. 168. Oxford University Press, Oxford (2017)
2. Pages, L., et al.: At. Data **4**, 1 (1972)
3. Particle Data Group, Tanabashi, M., et al.: Phys. Rev. **D98**, 030001 (2018)
4. Rossi, B.: High-Energy Particles. Prentice Hall, Englewood Cliffs, N.J. (1952)
5. Lohmann, W., Kopp, R., Voss, R.: Energy loss of muons in the energy range 1 GeV to 10 TeV, CERN yellow report, CERN 85–03. Genève, Switzerland (1985)
6. Kopp, R., et al.: Z. Phys. C **28**, 171 (1985)
7. Dietrich, S.S., Berman, B.L.: At. Data Nucl. Data Tables **38**, 199 (1988)
8. Rudstam, G.: Z. Naturforsch. **21a**, 1027 (1966)
9. Janni, J.F.: At. Data Nucl. Data Tables **27**, 147 (1982)
10. Wigmans, R.: Rev. Sci. Instr. **69**, 3723 (1998)
11. Leroy, C., Sirois, Y., Wigmans, R.: Nucl. Instr. Methods **A252**, 4 (1986)
12. Lachkar, J., et al.: Nucl. Phys. A **222**, 333 (1974)

# Chapter 3
# Shower Development

## 3.1 Electromagnetic Showers

In the absorption process of high-energy particles, the fundamental interactions described in the previous chapter are repeated many times. For example, a primary, multi-GeV electron may radiate thousands of photons on its way through the detector. The overwhelming majority of these photons are very soft, and are absorbed through Compton scattering and the photoelectric effect. The photons carrying more energy than 5–10 MeV may create $e^+e^-$ pairs. The fast electrons and positrons generated in these processes may in turn lose their energy by radiating more photons, which may create more electron–positron pairs, and so forth. The result is a shower that may consist of thousands of different particles: electrons, positrons and photons.

The shower energy is deposited by the numerous electrons and positrons, which ionize the atoms of the absorber material. Because of the multiplication mechanism described above, the number of electrons and positrons, and thus the amount of energy deposited in an absorber slice of given thickness, initially increases as the shower develops, i.e., with increasing shower depth.

However, as the shower develops, the average energy of the shower particles decreases, and at some point no further multiplication takes place. The depth at which this occurs is called the *shower maximum*. Beyond this depth, the shower photons are, on average, more likely to produce *one* electron in their (Compton or photoelectric) interactions, rather than an electron–positron *pair*. And the electrons and positrons are, again on average, more likely to lose their energy through ionization of the absorber medium, rather than by producing *new* photons (which would in turn convert into *more* electrons) through radiation. Beyond the shower maximum, the number of shower particles, and thus the energy deposited in a detector slice of given thickness, therefore gradually decreases.

All these aspects are illustrated in Fig. 3.1, which shows the energy deposit as a function of depth, for 1, 10, 100 and 1,000 GeV electron showers developing in a block of copper. The higher the initial energy of the showering particle, the longer the particle multiplication phase continues. The shower maximum is reached after about

© Springer Nature Switzerland AG 2019
M. Livan and R. Wigmans, *Calorimetry for Collider Physics, an Introduction*,
UNITEXT for Physics, https://doi.org/10.1007/978-3-030-23653-3_3

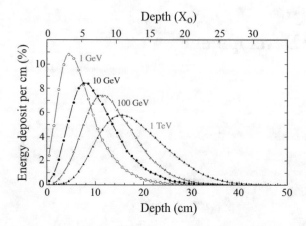

**Fig. 3.1** The energy deposit as a function of depth, for 1, 10, 100 and 1,000 GeV electron showers developing in a block of copper. In order to compare the energy deposit profiles, the integrals of these curves have been normalized to the same value. The vertical scale gives the energy deposit per cm of copper, as a percentage of the energy of the showering particle. Results of EGS4 calculations [1]

5 cm for 1 GeV electrons. For every order of magnitude in energy, this maximum shifts by ∼3.5 cm deeper into the detector, to reach a depth of about 16 cm for electron showers of 1 TeV. The copper thickness needed to absorb 99% of the shower energy rises from 23 cm at 1 GeV, via 28 cm at 10 GeV and 33 cm at 100 GeV, to 39 cm at 1 TeV.

### 3.1.1 Differences Between Electron and γ Induced Showers

Electron and $\gamma$ induced showers are different in the way they start developing (Fig. 3.2). Upon entering an absorber medium, electrons start to radiate immediately. In the first radiation length, they lose, on average, 63.2% of their kinetic energy to bremsstrahlung. On the other hand, high-energy $\gamma$s travel, on average, 9/7 radiation lengths in the absorbing medium before their first interaction.

This difference between the interaction mechanisms has two types of consequences for the em showers initiated by photons and by electrons/positrons:

1. Photon-induced showers deposit their energy, on average, deeper inside the absorbing structure than do em showers induced by charged particles of the same energy.
2. The fluctuations in the amount of energy deposited in a given slab of material are larger for showers induced by photons than for showers induced by $e^+$ or $e^-$.

The first effect results from the fact that the photons travel a certain distance in the absorbing structure before they start losing energy, while electrons and positrons

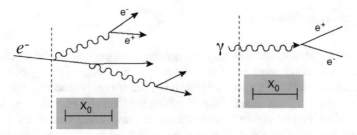

**Fig. 3.2** The different starts of em showers initiated by high-energy electrons (*left*) and $\gamma$s (*right*) entering an absorbing medium [1]

**Fig. 3.3** Distribution of the energy fraction deposited in the first five radiation lengths by 10 GeV electrons and $\gamma$s showering in lead. Results of EGS4 simulations. From: Wigmans, R. and Zeyrek, M. (2002). *Nucl. Instr. and Meth.* **A485**, 385

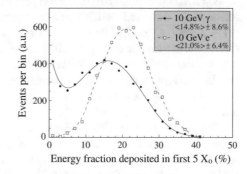

start losing energy immediately upon their entry. Moreover, the starting point of the photon-induced showers fluctuates from event to event, which leads to the second effect.

These effects are illustrated in Fig. 3.3, which shows the distribution of the energy deposited by 10 GeV electrons and 10 GeV photons in a $5X_0$ (2.8 cm) thick slab of lead [2]. On average, the electrons deposit more energy in this material than the photons (2.10 GeV vs. 1.48 GeV). However, the *fluctuations* in the energy deposited by the photons are clearly larger than those in the energy deposited by the electrons (0.86 GeV vs. 0.64 GeV). The distribution for the photon showers exhibits an excess near zero, which is the result of photons penetrating (almost) the entire slab without interacting. The "punch-thru" probability for a high-energy $\gamma$ is in this example $\exp(-35/9) \approx 2\%$.

The different effects of dead material installed in front of the calorimeter on electrons/positrons and $\gamma$s is particularly relevant for the ATLAS experiment, where the electromagnetic calorimeter is "hidden" in a cryostat, although the effects are less dramatic than suggested in Fig. 3.3 because this cryostat is made of aluminium. However, another consequence of the differences between electron and $\gamma$ induced showers for ATLAS is that the very complicated calibration scheme that was developed for electrons showering in the three longitudinal segments of the electromagnetic calorimeter is *not* necessarily the optimal solution for $\gamma$ detection in this calorimeter. More on this aspect in Chap. 9.

## 3.2 Hadron Showers

Just like in showers initiated by high-energy electrons or photons, the absorption of a high-energy hadron takes place in a sequence of processes of the types described in Sect. 2.3. In hadronic shower development, the strong interactions between the shower particles and the nuclei of the absorbing medium also play an important role and, therefore, hadron showers are much more complicated than electromagnetic ones. The variety of processes that may occur, both those at the particle level and those involving the struck nucleus, is much larger. This is true for all stages of the shower development.

When a high-energy charged hadron penetrates a block of matter, it will ionize the atoms of the traversed medium, in a continuous stream of events. It loses typically $\sim 1 - 2$ MeV $g^{-1}cm^2$ of its kinetic energy in this process. However, if the hadron carries enough kinetic energy, it will at some point encounter an atomic nucleus with which it interacts strongly. In this reaction, the hadron typically creates a number of new hadrons (primarily pions) and leaves the struck nucleus in a highly excited state, from which the latter decays by "boiling off" several nucleons or nucleon aggregates, followed by the emission of $\gamma$ rays.

The average distance a high-energy proton or neutron has to travel before a nuclear interaction is called the *nuclear interaction length*, $\lambda_{int}$. Since mesons, such as pions or kaons, are smaller, they are less likely to encounter a nucleus. They travel typically a 25% longer distance (and therefore lose 25% more energy) than protons before a nuclear interaction occurs [3]. In Sect. 3.3.3, $\lambda_{int}$ is discussed in more detail.

Neutral hadrons do not ionize the traversed medium. Nuclear reactions are their *only* option for losing energy. This is in particular true for neutrons, which are abundantly produced in hadronic shower development. As described in Sect. 2.3.3, neutrons deposit their kinetic energy in ways very different from those for the charged shower particles, with potentially very important implications for calorimetry.

The particles produced in the first nuclear reaction (mesons, nucleons, $\gamma$s, etc.) may in turn lose their energy by ionizing the medium and/or induce new (nuclear) reactions, thus causing a shower to develop. Conceptually, this shower is very similar to the em ones discussed in Sect. 2.1. Initially, the number of shower particles increases as a result of multiplication processes, and so does the energy deposited by the shower particles in a slice of given thickness. However, at some depth further multiplication is balanced by the absorption of shower particles. Beyond this shower maximum, the number of shower particles and the energy they deposit in a slice of matter of given thickness gradually decrease.

Despite these similarities, there are also major differences between em and hadronic showers. One of these differences concerns the *scale* of the shower development. Since hadronic shower development is governed by nuclear, instead of electromagnetic, interactions, the relevant scale is in this case the mean free path hadrons travel before initiating a reaction with an atomic nucleus, i.e., the nuclear interaction length $\lambda_{int}$. Figure 3.4 shows the *average* energy deposit as a function of depth for pions of different energies absorbed in the $^{238}$U/plastic-scintillator calorimeter of the

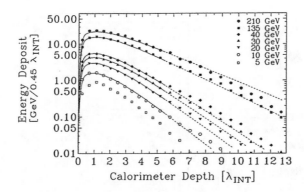

**Fig. 3.4** The energy deposit as a function of depth, for pion showers of different energies developing in the ZEUS $^{238}$U/plastic-scintillator calorimeter. These profiles are measured with respect to the starting point of the showers. From: Catanesi, M.G. et al. (1987). *Nucl. Instr. and Meth.* **A260**, 43

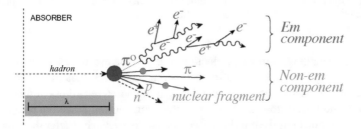

**Fig. 3.5** Schematic depiction of a hadron shower. The energy carried by the hadron is typically deposited in the form of an electromagnetic and a non-electromagnetic component. The em component is the result of $\pi^0$s and $\eta$s produced in the nuclear reactions. The non-em component consists of charged hadrons, and nuclear fragments. Some fraction of the energy transferred to this component (the "invisible" energy needed to break apart nuclei excited in this process) does *not* contribute to the calorimeter signals [1]

ZEUS experiment [4]. In stark contrast with the profiles for em showers (Fig. 3.1), the depth at which the hadronic profiles reach their maximum value barely changes with the energy of the incoming particles. This maximum is invariably located around $1\lambda_{int}$ beyond the starting point of the showers. However, the downward slope of the absorption profile clearly becomes less steep as the hadron energy increases.

The reason for the fact that the hadronic shower maximum is (almost) independent of the hadron energy is illustrated by Fig. 3.5, which schematically depicts the start of a typical high-energy hadron shower. In the first interaction, some fraction of the energy is used for the production of one or several $\pi^0$s. These decay into $\gamma$s and develop electromagnetic showers. This em shower component is a typical feature of hadron showers. On average, one third of the energy of the incoming high-energy hadron is carried away by $\pi^0$s. In $^{238}$U, the absorber material for which the results shown in Fig. 3.4 were obtained, the radiation length is 3.2 mm. One nuclear interaction length (100 mm) is thus equivalent to about $30X_0$. As illustrated by Fig. 3.1, the

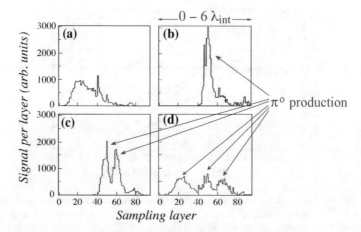

**Fig. 3.6** Longitudinal profiles for four different showers induced by 270 GeV pions in a lead/iron/plastic-scintillator calorimeter. Data from [5]

em showers resulting from the production of $\pi^0$s in the first nuclear interaction have thus completely been absorbed after the hadron shower has penetrated the absorber material over a distance of $1\lambda_{int}$ beyond its starting point, regardless of the energy of these $\pi^0$s and thus of the energy of the showering hadron. Since the relevant scale for the development of the em shower component is so much shorter than that of the non-em component, the energy deposit profile of the absorbed hadron is in its first stage completely determined by the em shower component. Therefore, the depth at which the energy deposited in a slice of given thickness reaches its maximum value is independent of the hadron energy.

It is crucial to note that the absorption profiles shown in Fig. 3.4 are *averaged* over large numbers of showers. The energy deposit profiles of *individual* hadron showers may deviate substantially from these averages. In individual showers, the production of energetic $\pi^0$s may occur at later stages of the shower development, i.e., initiated by secondary or tertiary shower particles. The em shower components created by such $\pi^0$s develop in that case in completely different regions of the absorbing volume, and this will lead to energy deposit profiles that differ considerably from the average. Figure 3.6 shows a few examples of non-average energy deposit profiles, for pion showers developing in a lead/iron/plastic-scintillator sandwich calorimeter [5]. The absorber structure consisted of 40 lead plates (3.1 mm thick), followed by 26 iron plates (2.5 cm thick). The energy deposited in this structure was measured in every individual scintillator plate, so that detailed event-by-event information on the longitudinal shower development was obtained.

Figure 3.6a shows a shower with an energy deposit profile that corresponds roughly with the profile averaged over a large number of showers. However, individual beam particles may penetrate deep into the detector before initiating a nuclear reaction (Fig. 3.6b, c). Figure 3.6b depicts an event in which a large fraction of the energy was transferred to one or several $\pi^0$s in this first nuclear interaction. Also in the event shown in Fig. 3.6c, energetic $\pi^0$s were produced in the first nuclear interaction. However, in addition, an energetic charged hadron was produced at that point. This particle traveled about one interaction length deeper into the absorber and then transferred almost all its energy to $\pi^0$s in a second-generation interaction, leading to a two-peak structure in the longitudinal energy deposit profile. In Fig. 3.6d, even three generations of $\pi^0$ production can be distinguished.

Such "stochastic" energy deposit profiles are by no means exceptional in hadronic shower development. They are a consequence of the fact that the $\pi^0$s produced in the shower development deposit the energy they carry in a much smaller absorber volume than other shower particles. Therefore, the hadronic energy deposit profiles directly reflect the large event-by-event fluctuations that may occur in both the energy carried by these $\pi^0$s and the position in the absorber where they are generated.

## 3.3  Material Dependence

The large difference between the values of the radiation length and the nuclear interaction length, which is responsible for the phenomena discussed in Sect. 3.2 (Figs. 3.4 and 3.6), is a consequence of the difference between the cross sections for the reactions mediated by the em and the strong interactions. These depend on the $A$ and $Z$ values of the absorbing medium. In the following subsections, this dependence is described.

### 3.3.1  The Radiation Length

Since the em shower development is primarily determined by the electron density in the absorber medium, it is to some extent possible, and in any case convenient, to describe the shower characteristics in a material-independent way. It can be shown that the asymptotic cross section for *photon interactions* is related to $X_0$ as

$$\sigma(E \to \infty) = \frac{7}{9} \frac{A}{N_A X_0} \tag{3.1}$$

in which $X_0$ is expressed in g cm$^{-2}$ and the ratio of Avogadro's number ($N_A$) and the atomic weight ($A$) denotes the number of atoms per gram of material. This implies that the mean free path of very-high-energy photons equals $\frac{9}{7} X_0$. For approximate calculations, which are accurate to within 3%, the Particle Data Group [6] recommends the following expression:

$$X_0 = \frac{716.4 \, A}{Z \, (Z+1) \, \ln{(287/\sqrt{Z})}} \ \text{g cm}^{-2} \tag{3.2}$$

The radiation length for a mixture of different materials can be calculated as follows

$$\frac{1}{X_0} = \sum_i V_i / X_i \tag{3.3}$$

in which $V_i$ and $X_i$ are the fraction by volume and the radiation length (expressed in mm) of the $i$th component of the mixture. Equation 3.3 may, for example, be used to calculate the effective radiation length of a calorimeter consisting of a variety of different materials. Let us, as an example, consider a lead/liquid-argon calorimeter consisting of 5 mm thick lead plates, separated by 3 mm wide LAr-filled gaps. The radiation lengths of lead and LAr are 5.6 mm and 140 mm, respectively, and the fractional volume occupied by these elements is 62.5% for lead and 37.5% for argon. Therefore, we find for the *effective radiation length*: $X_{\text{eff}} = [0.625/5.6 + 0.375/140]^{-1} = 8.75$ mm, only slightly less than the value one would obtain if the argon were replaced by vacuum ($X_{\text{eff}} = 5.6/0.625 = 8.96$ mm).

If the argon were contained in 0.8 mm thick stainless steel containers, separated from the lead plates by 0.2 mm of air, the volume ratio of the different materials in the calorimeter structure would be as follows: lead/argon/iron/air = 5/3/1.6/0.4 = 50%/30%/16%/4%. With the radiation lengths for iron and air being 17.6 mm and 300 m, respectively, the effective radiation length of this structure becomes: $X_{\text{eff}} = [0.5/5.6 + 0.3/140 + 0.16/17.6 + 0.04/30, 000]^{-1} = 9.95$ mm.

The radiation length of a *compound* can be calculated in a similar way, using the equation

$$\frac{1}{X_0} = \sum_i m_i / X_i \tag{3.4}$$

in which $m_i$ and $X_i$ are the fraction (by mass) and the radiation length (expressed in g cm$^{-2}$) of the $i$th component of the compound. Let us, for example, calculate the radiation length of lead-tungstate crystals (PbWO$_4$). The mass ratio of the elements of which these crystals are composed is as follows: lead/tungsten/oxygen = 207.19/183.85/64.0 = 45.5%/40.4%/14.1%. The radiation lengths of these elements are 6.37, 6.76 and 34.24 g cm$^{-2}$, respectively. Therefore, we find for the radiation length of lead-tungstate: $X_0 = [0.455/6.37 + 0.404/6.76 + 0.141/34.24]^{-1} = 7.39$ g cm$^{-2}$. Since the density of these crystals amounts to 8.30 g cm$^{-3}$, their radiation length equals 8.9 mm.

### 3.3.2   The Molière Radius

The Molière radius is frequently used to describe the transverse development of em showers in an *approximately* material independent way. This quantity does not have a physics meaning equal in precision to that of the radiation length, which describes the longitudinal development. It is defined in terms of the radiation length $X_0$ and the critical energy $\varepsilon_c$ (Sect. 2.2.1), as follows:

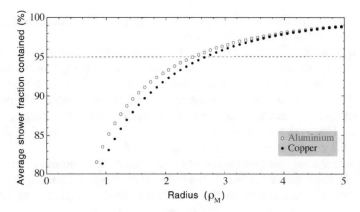

**Fig. 3.7** Average energy fraction contained in an infinitely long cylinder of absorber material, as a function of the radius of this cylinder. Results of EGS4 calculations for various absorber materials and different energies [1]

$$\rho_M = E_s \frac{X_0}{\varepsilon_c} \tag{3.5}$$

in which the scale energy $E_s$, defined as $m_e c^2 \sqrt{4\pi/\alpha}$, equals 21.2 MeV. Typically, $\sim$85–90% of the shower energy is deposited in a cylinder with radius $\rho_M$ around the shower axis (Fig. 3.7).

The Molière radii for mixtures or compounds of different elements may be calculated in the same way as the radiation length for such mixtures or compounds was obtained, replacing $X_i$ in Eqs. 3.3 and 3.4 by $\rho_i$. Let us, for example, calculate the Molière radius of BGO crystals, which have the following chemical composition: $Bi_3Ge_4O_{12}$. The mass ratio of the elements of which these crystals are composed is as follows: bismuth/germanium/oxygen = $(209.0\times3)/(72.6\times4)/(16.0\times12)$ = 56.5%/26.2%/17.3%. The radiation lengths of these elements are 6.32, 12.25 and 34.24 g cm$^{-2}$, respectively. For the critical energies, we use Eq. 2.3, which gives values of 7.24 MeV for bismuth, 18.4 MeV for germanium and 66 MeV for oxygen. This leads to Molière radii for the different crystal components of 18.5 g cm$^{-2}$ (Bi), 14.4 g cm$^{-2}$ (Ge) and 11.0 g cm$^{-2}$ (O), respectively. When combining all these data, we find for the Molière radius of bismuth-germanium oxide: $\rho_M = [0.565/18.5 + 0.262/14.4 + 0.173/11.0]^{-1} = 15.5$ g cm$^{-2}$. Since the density of these crystals amounts to 7.13 g cm$^{-3}$, their Molière radius thus equals about 22 mm.

The Molière radius is much less $Z$ dependent than the radiation length. This can be seen from Eqs. 2.3, 3.2 and 3.5. The radiation length scales in first approximation with $A/Z^2$ (Eq. 3.2). If we assume that $A$ is proportional to $Z$, which is roughly true, the radiation length (expressed in g cm$^{-2}$) decreases with increasing $Z$ like $1/Z$. The same is approximately true for the critical energy (Eq. 2.3). Since the Molière radius is defined as the *ratio* of the radiation length and the critical energy, the $Z$ dependence cancels in first approximation.

This difference in $Z$ dependence may be illustrated by comparing two materials that are frequently used as absorbers in calorimeters, copper ($Z = 29$) and lead ($Z = 82$). The densities are not very different: 8.96 g cm$^{-3}$ for copper versus 11.35 g cm$^{-3}$ for lead. The radiation lengths for these materials reflect the large difference in $Z$: 14.3 mm for copper versus 5.6 mm for lead, almost a factor of three difference. However, the values for the Molière radii show a completely different pattern: 15.2 mm for copper versus 16.0 mm for lead.

As a consequence, the development of em showers in these two absorber materials has very different characteristics. In the longitudinal direction, it takes about three times as much copper as lead (in cm) to contain these showers. However, laterally, the showers in copper are even *narrower* than those in lead.

### 3.3.3  The Nuclear Interaction Length

The nuclear interaction length of an absorber medium is defined as the average distance a high-energy hadron has to travel inside that medium before a nuclear interaction occurs. The probability that the particle traverses a distance $z$ in this medium *without* causing a nuclear interaction equals

$$P = \exp\left(-z/\lambda_{\text{int}}\right) \qquad (3.6)$$

This definition is thus equivalent to the one for the mean free path of high-energy photons, which was found to be equal to 9/7 of a radiation length (Eq. 3.1). And just as the mean free path of photons is inversely proportional to the total cross section for photon-induced reactions, $\lambda_{\text{int}}$ is inversely proportional to the total cross section for nuclear interactions:

$$\sigma_{\text{tot}} = \frac{A}{N_A \lambda_{\text{int}}} \qquad (3.7)$$

This cross section is determined by the size of the projectiles and the size of the target nuclei. The cross section of the target nuclei is determined by their radius squared. And since the volume of these nuclei (and thus $r^3$) scales with the atomic weight $A$, the cross section scales with $A^{2/3}$. From Eq. 3.7, it then follows that $\lambda_{\text{int}}$ scales with $A^{1/3}$, when expressed in units of g cm$^{-2}$ (which eliminates differences in material density).

The smallest values for $\lambda_{\text{int}}$, around 10 cm, are found for high-density, high-$Z$ materials such as tungsten, gold, platinum and uranium. For frequently used absorber materials such as iron and copper, the interaction length is less than twice as long (a 60–70% increase compared with uranium). This is quite different from the situation encountered earlier for the radiation length, which increases by about a factor of five going from uranium to iron. We will come back to these differences when discussing calorimetric particle identification (Sect. 11.3).

The nuclear interaction lengths for mixtures of different elements or for a compound can be determined in the same way as discussed for the radiation length and the Molière radius (Sects. 3.3.1 and 3.3.2).

### 3.3.4 The Ratio $\lambda_{int}/X_0$

In Sect. 3.2, we saw that, in uranium, the nuclear interaction is more than 30 times larger than the radiation length. Given the $Z$ and $A$ dependence of these scaling variables, the ratio $\lambda_{int}/X_0$ depends on these material parameters as $Z^2 A^{-2/3}$, and thus increases (almost proportionally) with the $Z$ value of the absorber material.

Figure 3.8 shows the actual value of this ratio as a function of the $Z$ of some absorber materials that may be used in calorimeters. It increases from 4.4 for aluminium ($Z = 13$) and 9.5 for iron ($Z = 26$) to 27 for tungsten ($Z = 74$ and 33 for uranium ($Z = 92$). The large difference between $X_0$ and $\lambda_{int}$ makes high-$Z$ absorber materials ideally suited for particle identification. When a beam consisting of a mixture of high-energy electrons and pions is sent through a 5 mm thick plate of lead, almost all (96%) of the pions will traverse this plate without strongly interacting. On the other hand, the electrons lose a considerable fraction of their energy by radiating large numbers of bremsstrahlung photons. This feature is the founding principle of some *preshower detectors* (Sect. 5.4).

**Fig. 3.8** Ratio of the nuclear interaction length and the radiation length as a function of the $Z$ value of the absorber material [1]

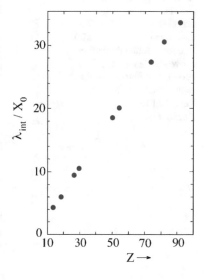

## 3.4 The Importance of the Last Stages of the Absorption Process

Eventually, all the available energy of the incoming particle is shared among a very large number of shower particles, each of which carries very little kinetic energy, and that energy is transferred to the atoms and molecules of the absorber medium. And even though calorimeters are often intended to measure energy deposits at the level of $10^9$ eV and up, their performance is in practice determined by what happens at the MeV, keV and sometimes eV levels, simply because the particles that carry these low energies are so abundantly produced in the absorption process.

To illustrate the importance of the last stages of the shower development, Fig. 3.9 shows what happens to the energy carried by a 10 GeV electron that is absorbed in a block of material, as a function of the $Z$-value of that material. If we take as an example lead, which is often used as absorber material in calorimeters intended for detecting such electrons, it turns out that almost 70% of the entire energy is deposited by shower electrons that carry less than 4 MeV kinetic energy. Electrons of less than 1 MeV even carry 40% of the total energy, which means that there must thus be *at least* 4,000 such electrons, and probably many more. Figure 2.5 shows that these soft electrons are almost exclusively the result of Compton scattering and the

**Fig. 3.9** The composition of em showers. Shown are the percentages of the energy of 10 GeV electromagnetic showers deposited through shower particles with energies below 1 MeV (the dashed curve), below 4 MeV (the dash-dotted curved) or above 20 MeV (the solid curve), as a function of the $Z$ of the absorber material. Results of EGS4 simulations [1]

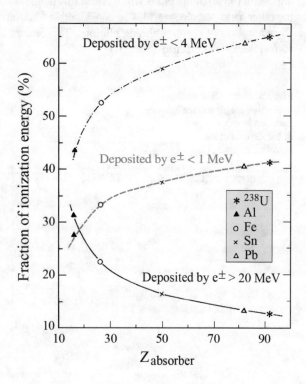

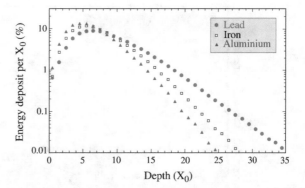

**Fig. 3.10** Energy deposit as a function of depth, expressed in radiation lengths, for 10 GeV electron showers developing in aluminium, iron and lead. Results of EGS4 calculations [1]

photoelectric effect, since these processes are the most likely ones to occur when $\gamma$s with energies below 4 MeV interact with the absorber material.

To understand the peculiarities of electromagnetic calorimeters, one thus needs to understand the relevant characteristics of Compton scattering and the photoelectric effect. We mention three examples.

1. The *radiation length* ($X_0$) was introduced as a parameter to describe electromagnetic shower development in a material independent way. Longitudinal shower profiles should thus be material independent when expressed in terms of $X_0$. Figure 3.10 shows that this is by no means the case.

   Figure 5.1a also illustrates the inadequacy of this parameter for determining the calorimeter depth needed to contain em showers at the level of 99%. For 10 GeV electrons, this depth ranges from $16X_0$ to $22X_0$, depending on the chosen absorber material. These discrepancies are a result of the fact that $X_0$ is defined based on the properties of the high-energy component of the showers, where electrons predominantly lose energy by radiation (Bremsstrahlung) and $\gamma$s interact by producing $e^+e^-$ pairs. These processes do not play an important role in the last stages of the shower development during which, as shown above, a major fraction of the total energy is deposited.

2. There is no preferential direction for the electrons produced in Compton scattering and by the photoelectric effect. Because of the dominant role of these processes in em shower development, a large fraction of *the shower particles thus travel in random directions* with respect to the particle that initiated the shower and whose properties are being measured with the calorimeter (Fig. 3.11). This means that the orientation of the active layers of a sampling calorimeter can be chosen as desired, without serious implications for the calorimetric performance of the detector.

   The first generation of sampling calorimeters used in particle physics experiments consisted almost exclusively of instruments of the "sandwich type," i.e., detectors composed of alternating layers of absorber and active material, oriented perpendicular to the direction of the particles to be detected. Although this, from an intuitive point of view, may seem to be the only correct choice, the R&D with

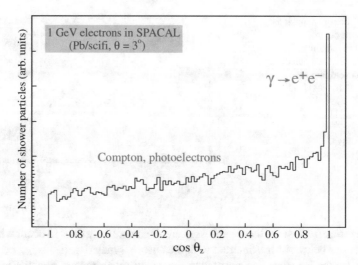

**Fig. 3.11** Angular distribution of the shower particles (electrons and positrons) through which the energy of a 1 GeV electron is absorbed in a lead-based fiber calorimeter. The angle between the direction of the shower particles and the fiber axis ($\theta_z$) was chosen to be 3° in these EGS4 Monte Carlo simulations. From: Acosta, D. et al. (1990). *Nucl. Instr. and Meth.* **A294**, 193

fiber calorimeters has proven that there is no need for such a geometry [7, 8]. The notion that the active calorimeter layers do not necessarily have to be oriented perpendicular to the direction of the incoming particles has had a considerable impact on the design of detectors for new experiments (Fig. 1.11). Other orientations may offer considerable advantages in terms of detector hermeticity, readout, granularity, etc. Apart from the "spaghetti" type of calorimeters built for a number of experiments, this development is also illustrated by the liquid-argon calorimeters with an "accordion" geometry and the tile/fiber hadron calorimeter of the ATLAS experiment at the LHC [9].

3. *The cross section for the photoelectric effect is extremely Z-dependent ($\propto Z^5$).* This has very important consequences for sampling calorimeters that consist of high-$Z$ absorber material and low-$Z$ active layers. Low-energy $\gamma$s produced in the shower development will, for all practical purposes, **only** interact in the absorber material, and the photoelectrons produced in this process will only contribute to the signals if they are produced very close to a boundary layer. In practice, they are much less efficiently sampled than the high-energy electron/positron pairs produced in the early stages of the shower development. As a result, the sampling fraction in such calorimeters decreases as the shower develops in depth (see Fig. 3.13). Also, the fact that $e/mip \neq 1$ in such calorimeters[1] is the result of this phenomenon. Some consequences for calorimetry are discussed in Sect. 7.3.

---

[1] See Sect. 7.1 for the definition of $e/mip$.

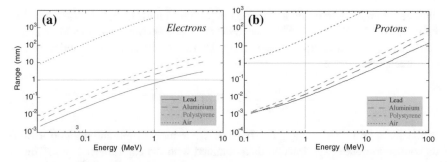

**Fig. 3.12** Average range of electrons (**a**) and protons (**b**) in various absorber materials, as a function of energy. Data from [10]

The fact that a large fraction of the em shower energy is deposited by very soft electrons also has implications for the energy resolution of em sampling calorimeters. This may be concluded from Fig. 3.12a, which shows the range of electrons with energies from 20 keV to 10 MeV in several materials. For example, 1 mm of aluminium is enough to stop electrons with a kinetic energy up to 590 keV. For silicon, a material that is increasingly being considered as active material for sampling calorimeters, electrons with energies up to 115 keV stop in 100 $\mu$m thick layers, while 330 keV electrons have a range of 500 $\mu$m [10]. If one replaced in the same absorber structure 500 $\mu$m thick Si sensors by 100 $\mu$m thick ones, shower electrons that escaped from the absorber layers carrying a kinetic energy between 115 keV and 330 keV would thus always produce the same signals in the thicker sensors, while the signals in the thinner ones would depend on the angle of incidence. Because of this additional source of fluctuations, the energy resolution would deteriorate as the sensors were made thinner and thinner. The fact that this does indeed happen (Sect. 4.2) is an indication of the important contributions of shower electrons from this very-low-energy bracket.

## 3.5 The Sampling Fraction

### 3.5.1 Minimum Ionizing Particles

An important parameter characterizing sampling calorimeters is the *sampling fraction*. We define the sampling fraction of a calorimeter as the energy deposited *by minimum ionizing particles* in the active calorimeter layers, measured relative to the *total energy* deposited by such particles in the calorimeter. Let us consider, as an example, the D0 calorimeter [11] which operated for more than 20 years at Fermilab's Tevatron collider. The hadronic section of this calorimeter consisted of 3 mm thick depleted uranium plates ($^{238}$U) interleaved with 5 mm wide gaps filled with liquid argon. Minimum ionizing particles lose, on average, 2.13 MeV cm$^{-1}$ in LAr and 20.5 MeV cm$^{-1}$ in $^{238}$U (see Table B.1), or 1.06 MeV and 6.15 MeV in one active and one passive layer, respectively. Therefore, the sampling fraction of this calorimeter amounts to

$$\frac{1.06}{1.06 + 6.15} \; = \; 0.147, \text{ or } 14.7\%$$

While this definition of the sampling fraction is extremely simple and straightforward, its connection with experimental data obtained with the calorimeter is not. This is because the response to mips cannot be measured directly. It should be emphasized that a mip is a *hypothetical particle*. As soon as a charged particle with an energy for which $dE/dx$ reaches its minimum value starts traveling through matter, it loses energy and therefore ceases to be a mip.

For all practical purposes, muons are the closest thing nature provides us with in terms of mips. However, even muons with an energy as low as 5 GeV are by no means minimum ionizing particles. Since they are extremely relativistic ($\gamma \sim 50$), the energy loss per unit length is significantly larger than the minimum ionizing value [12]. The increased specific ionization of muons with energies larger than the minimum ionizing value is due to phenomena such as $\delta$-ray emission (relativistic rise), bremsstrahlung, $e^+e^-$ pair production and, at very high energies, nuclear reactions. The contribution of these effects to the total energy loss is strongly dependent on the muon energy and on (the $Z$ value of) the traversed material.

In practice, the experimental calorimeter response to mips is determined from the signal distributions measured for muons of different energies, by estimating the consequences of the above effects, thereby unfolding the *mip* part of the calorimeter signals [13–16].

### 3.5.2  Sampling of Non-mips

The sampling fraction of a calorimeter, while defined for minimum ionizing particles, may be very different for other types of particles. This may have very large and important consequences for the characteristics of sampling calorimeters, such as the response function for different types of high-energy particles, and the energy resolution with which these can be measured. In this context, we discuss the two most important effects in that context.

1. *Soft $\gamma$s.*

   In Sect. 2.2.3, we saw that soft $\gamma$s produced in em shower development predominantly interact with the absorber material through the photoelectric effect (Fig. 2.5). The cross section for this process depends very sensitively on the $Z$ value. As pointed out in Sect. 3.4, this has important consequences for sampling calorimeters that consist of high-$Z$ absorber material and low-$Z$ active layers. Low-energy $\gamma$s are sampled much less efficiently than mips in such calorimeters.

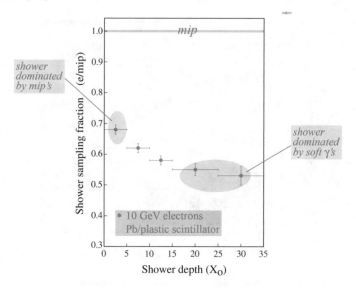

**Fig. 3.13** The sampling fraction changes in a developing shower. The local sampling fraction, normalized to that for a minimum ionizing particle, is shown as a function of depth for 10 GeV electrons in a Pb/scintillating-plastic calorimeter. Results of EGS4 calculations [1]

The composition of an em shower changes as a function of depth, or age. In the late stages, most of the energy is deposited by soft $\gamma$s which undergo Compton scattering or photoelectric absorption, and the sampling fraction for this shower component (i.e., the fraction of the deposited energy that contributes to the calorimeter signals) may be very different from that of the mip-like $e^+e^-$ pairs that dominate the early stages of the shower development.

Figure 3.13 illustrates how the sampling fraction of a given calorimeter structure depends on the stage of the developing showers. In calorimeters consisting of high-$Z$ absorber material (in this case, lead) and low-$Z$ active material (in this case, plastic), the sampling fraction may vary by as much as 25–30% over the volume in which the absorption takes place [17]. A particular problem resulting from this phenomenon concerns the intercalibration of the different sections of a longitudinally segmented calorimeter system. Since the longitudinal size of an em shower and the depth at which the shower maximum is reached depend on the energy (Fig. 3.1), and are also different for electrons and $\gamma$s of the same energy (Fig. 3.2), this phenomenon complicates the conversion of the measured signals from a longitudinally segmented calorimeter into deposited energy. This is discussed in detail in Chap. 9.

2. *Slow neutrons.*

In Sect. 2.3.3, we saw that neutrons in the energy range 1 eV–1 MeV for all practical purposes only interact with the absorber material through elastic scattering. In hadronic shower development, typically, 5–10% of the energy is carried by evaporation neutrons. In practice, most of the kinetic energy carried by these

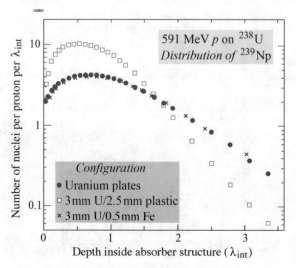

**Fig. 3.14** The longitudinal distributions of $^{239}$Np nuclei (from neutron capture in $^{238}$U) produced by hadron showers in various calorimeter configurations. From: Leroy, C., Sirois, Y. and Wigmans, R. (1986). *Nucl. Instr. and Meth.* **A252**, 4

evaporation neutrons is deposited through elastic scattering. The reaction products of elastic scattering are a recoil nucleus and a lower-energy neutron. The efficiency of this process for slowing down the neutrons depends on the mass of the absorber nuclei: the smaller this mass, the larger the average energy transferred in this process. Hydrogen is thus by far the most efficient medium to absorb the kinetic energy carried by these soft neutrons and thus thermalize them.

The spectacular effects of adding hydrogen to a calorimeter structure become clear from Fig. 3.14. In this figure, the longitudinal distributions of radioactive $^{239}$Np nuclei, produced in the absorption of 591 MeV protons in various calorimeter configurations, are shown [18]. This nuclide is the decay product of $^{239}$U, which is produced by thermal neutron capture in the absorber material, $^{238}$U. These distributions are thus indicative for the longitudinal distributions of thermal neutrons produced in the absorption of 591 MeV protons in these structures. The data points represented by closed circles and crosses are almost indistinguishable in this figure. These data points describe the neutron distributions in a stack of 3 mm thick $^{238}$U plates and a stack in which these plates were interleaved by 0.5 mm thick iron foils, respectively.

However, the distribution was dramatically affected (note the logarithmic vertical scale) when these iron foils were replaced by plastic plates with an equivalent thickness, in nuclear interaction lengths. The open squares in Fig. 3.14 represent the data measured for this configuration. Clearly, the neutrons were much more rapidly thermalized in this case and, therefore, must have lost a considerable fraction of their kinetic energy in elastic collisions with hydrogen nuclei in the plastic. This conclusion was confirmed by the observation of a significant reduction

**Table 3.1** Sampling fractions for neutrons and mips in various calorimeter structures. See text for details

| Structure ↓ Particle → | 1 keV $n$ | 10 keV $n$ (%) | 100 keV $n$ (%) | 1 MeV $n$ (%) | mip (%) |
|---|---|---|---|---|---|
| Fe/LH$_2$ (1/1 vol.) | 93.6 | 95.9 | 95.6 | 92.6 | 2.4 |
| Pb/LH$_2$ (1/1 vol.) | 99.2 | 99.2 | 98.8 | 98.3 | 2.2 |
| Fe/LAr (1/1 vol.) | 2.0 | 2.8 | 11.4 | 21.5 | 15.7 |

(by ∼20%) in the number of nuclear fissions in this structure, compared with the pure $^{238}$U case. Nuclear fission of $^{238}$U requires a neutron with a kinetic energy of at least 1.5 MeV, and because of the moderating effect of the plastic, the number of neutrons capable of inducing fission was smaller in this experiment.

These experimental results show that neutrons lose a disproportionally large fraction of their kinetic energy in collisions with hydrogen when the calorimeter structure contains this material. In addition, the recoil protons produced in this process may also substantially contribute to the calorimeter signals. As a result, neutrons in the energy bracket between 1 keV and a few MeV are sampled very differently from charged particles in calorimeter structures that contain hydrogenous active components. This has spectacular calorimetric consequences.

Table 3.1 shows the sampling fractions for neutrons of different energies for three different structures: iron/liquid-hydrogen, lead/liquid-hydrogen and iron/liquid-argon, each with equal volumes of metal and liquid. These sampling fractions were calculated on the basis of the cross sections for elastic scattering and the assumption that the average fraction of the neutron's kinetic energy transferred in elastic collisions with a nucleus containing $A$ nucleons, $\langle f_{el} \rangle$, is given by

$$\langle f_{el} \rangle = \frac{2A}{(A+1)^2} \tag{3.8}$$

which implies $\langle f_{el} \rangle$ values of 50% and 5% for hydrogen and argon, respectively. Neutrons at energies between 1 keV and 1 MeV lose almost 100% of their kinetic energy in the hydrogen in the Fe/H$_2$ and Pb/H$_2$ structures, while mips deposit only a few percent of their energy in the hydrogen. Therefore, if hydrogen were the active component of these structures, the sampling fraction for neutrons would be larger than that for mips by a factor of 40–50 (i.e., $n/mip$ = 40–50). In the liquid-argon case, mips are typically sampled more efficiently than soft neutrons (i.e., $n/mip$ < 1).

The $n/mip$ ratio depends sensitively on the fraction of hydrogen contained in the calorimeter structure. This is illustrated in Table 3.2 for the case of lead/hydrogen. Whereas the sampling fraction for mips is almost proportional to the fraction of hydrogen in the structure, the sampling fraction for neutrons changes by less than a factor of three when the fraction of hydrogen is changed by two orders of magni-

**Table 3.2** Sampling fractions for 1 MeV neutrons and mips in Pb structures containing different fractions of liquid hydrogen, assumed to be the active calorimeter medium in these configurations

| H$_2$ fraction (vol. %) | 1 MeV neutrons (%) | mips (%) | $n/mip$ ratio |
|---|---|---|---|
| 1% | 36.9 | 0.0227 | 1630 |
| 5% | 75.3 | 0.118 | 640 |
| 10% | 86.6 | 0.249 | 350 |
| 20% | 93.5 | 0.558 | 170 |
| 50% (Pb/H$_2$ = 1/1) | 98.3 | 2.20 | 45 |
| 90% | 99.8 | 16.8 | 5.9 |

tude. As a result, the $n/mip$ ratio spans a wide domain, from 5.9 in hydrogen-rich structures (90 volume%) to 1630 in hydrogen-poor ones (1 volume %).

The neutrons are thus much more efficiently sampled than mips in these calorimeter structures, more so if the fraction of hydrogen is reduced. At first sight, it might seem counter-intuitive that it takes a *reduction* in the fraction of neutron-sensitive material to *increase* the relative calorimeter response to these particles (the $n/mip$ ratio). However, this is easily understood from the fact that a change in the hydrogen fraction affects the response to mips much stronger than it affects the response to neutrons. The mips share their energy among the active and passive materials according to the relative abundance of these materials in the structure. If there is a small fraction of active material, only a small fraction of the mip's energy is deposited in this material.

The neutrons, on the other hand, have in practice almost no alternative for depositing their kinetic energy in the active material, even if this active material represents only a small fraction of the total mass. This is because elastic scattering in the lead absorber is an extremely inefficient process for losing energy. On average, the neutrons lose only 0.96% of their kinetic energy in such collisions, versus 50% in hydrogen.

Apart from the effects discussed above, there are also differences between the sampling fractions for mips and non-mip charged shower particles. For example, the effects of the relativistic rise in Fig. 2.3 depends on the $Z$ value of the traversed material, and are thus in principle different for the active and passive layers of a sampling calorimeter. However, these effects are typically only at the level of a few percent. The differences may be considerably larger for extremely non-relativistic shower particles. However, the range of these particles is very short, so that they typically are not sampled, but rather range out in the material in which they are produced. More information about this issue is given in [1].

# References

1. Wigmans, R.: Calorimetry—Energy Measurement in Particle Physics, 2nd edn. International Series of Monographs on Physics, vol. 168. Oxford University Press, Oxford (2017)
2. Wigmans, R., Zeyrek, M.: Nucl. Instrum. Methods **A485**, 385 (2002)
3. Křivková, P., Leitner, R.: http://cds.cern.ch/record/683812/files/tilecal-99-007.pdf
4. Catanesi, M.G., et al.: Nucl. Instrum. Methods **A260**, 43 (1987)
5. Green, D.: In: Menzione, A., Scribano, A. (eds.), Proceedings of 4th International Conference on Calorimetry in High Energy Physics, La Biodola, Italy, p. 1. World Scientific, Singapore (1994)
6. Particle Data Group, Tanabashi, M., et al.: Phys. Rev. **D98**, 030001 (2018)
7. Acosta, D., et al.: Nucl. Instrum. Methods **A294**, 193 (1990)
8. Livan, M., Vercesi, V., Wigmans, R.: Scintillating-fibre Calorimetry, CERN Yellow Report, CERN 95–02, Genève, Switzerland (1995)
9. Aad, G., et al., ATLAS Collaboration: J. Instrum. **3**, S08003 (2008)
10. Berger, M.J.: Stopping Power and Range Tables for Electrons, Protons and Helium Ions, report NISTIR 4999 (1993). http://physics.nist.gov/Star
11. Abolins, M., et al.: Nucl. Instrum. Methods **A280**, 36 (1989)
12. Lohmann, W., Kopp, R., Voss, R.: Energy loss of muons in the energy range 1 GeV to 10 TeV, CERN yellow report, CERN 85–03, Genève, Switzerland (1985)
13. Åkesson, T., et al.: Nucl. Instrum. Methods **A262**, 243 (1987)
14. Bernardi, E., et al.: Nucl. Instrum. Methods **A262**, 229 (1987)
15. Acosta, D., et al.: Nucl. Instrum. Methods **A320**, 128 (1992)
16. Ajaltouni, Z., et al.: Nucl. Instrum. Methods **A388**, 64 (1997)
17. Wigmans, R.: On the calibration of segmented calorimeters. In: Proceedings of XII International Conference on Calorimetry in High Energy Physics, Chicago 2006, AIP Conference Proceedings, vol. 867, p. 90 (2006)
18. Leroy, C., Sirois, Y., Wigmans, R.: Nucl. Instrum. Methods **A252**, 4 (1986)

# Chapter 4
# The Calorimeter Signals

## 4.1 Introduction

Calorimeter signals consist typically of electric pulses produced either directly by collecting electric charge (e.g., in liquid-argon calorimeters or resistive plate chambers), or by photons that produce such pulses as a result of the photoelectric effect (e.g., in photomultiplier tubes, silicon photomultipliers or photodiodes). Since calorimeters are intended for measuring energy, the total signal should be a good measure of that energy. In practice, there are a number of *instrumental effects* that may spoil the relationship. These effects should be clearly distinguished from the *physics* effects described in the previous subsections (invisible energy, inefficient sampling of certain types of shower particles, or in late stages of the shower development, etc.).

## 4.2 Signal Linearity and Non-linearity

One of the most important characteristics of a calorimeter is *signal linearity*. This property is frequently the source of misconceptions, the most common one of which is that a calorimeter is linear if the average signals plotted versus the deposited energy can be described with a straight line. This is **not** what is meant by a linear calorimeter. The straight line has to extrapolate through the origin of the plot. Signal linearity means that the average calorimeter signal is *proportional* to the deposited energy.

Figure 4.1 illustrates this issue. The experimental data were obtained with a W/Si em calorimeter built by CALICE [1]. The authors fit the measured signals with the following expression:

$$E_{\text{mean}} = \beta \, E_{\text{beam}} - 360 \, \text{MeV} \tag{4.1}$$

Then, they define

© Springer Nature Switzerland AG 2019
M. Livan and R. Wigmans, *Calorimetry for Collider Physics, an Introduction*,
UNITEXT for Physics, https://doi.org/10.1007/978-3-030-23653-3_4

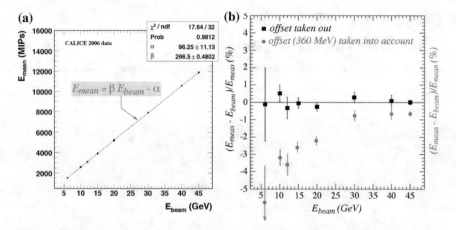

**Fig. 4.1** Average signal as a function of electron energy for the W/Si ECAL built by CALICE (**a**) [1]. Residual signals from this detector, before and after taking out a 360 MeV offset (**b**). From: Wigmans, R. (2018), *J. Progr. Part. Nucl. Phys.* **103**, 109

$$E_{\text{meas}} = E_{\text{mean}} + 360 \, \text{MeV} \tag{4.2}$$

and plot

$$(E_{\text{meas}} - E_{\text{beam}})/E_{\text{meas}}$$

as a function of the beam energy. The result is represented by the (black) squares in Fig. 4.1b. They conclude that "*the calorimeter is linear to within approximately 1%.*" This is highly misleading. When the calorimeter signals they actually *measured* are used to check the linearity, i.e., when

$$(E_{\text{mean}} - E_{\text{beam}})/E_{\text{mean}}$$

is plotted as a function of the beam energy, the results, represented by the (red) full circles in Fig. 4.1b, look quite different. We conclude from these results that the authors measured a signal non-linearity of 5% over one decade in energy.

Signal linearity is a very important property, since it is a crucial ingredient for the precision with which the energy of an unknown object that produces signals in the calorimeter can be measured. The assumption that it is sufficient to know the relationship between the signals and the deposited energy, even in the case of non-proportionality, constitutes another frequent misconception.

Let us take a look at Eq. 4.2, which describes the relationship between the measured signal ($E_{\text{mean}}$) and the corresponding energy ($E$)$_{\text{meas}}$ for this non-linear calorimeter. A $\gamma$ with an energy of 10 GeV would thus produce an average signal $E_{\text{mean}}$ of $10 - 0.36 = 9.64$ GeV. For a $\gamma$ of 5 GeV, the average signal would be $5 - 0.36 = 4.64$ GeV. Now consider what would happen to a 10 GeV $\pi^0$ that

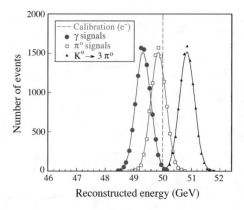

**Fig. 4.2** Signal distributions for $\gamma$s and various hadrons decaying into all-$\gamma$ final states. All particles have the same nominal energy and the detector, which consists of two longitudinal segments and has an intrinsic resolution of 0.5% for em showers of this energy, was calibrated with electrons such as to optimize the energy resolution for these particles. From: Wigmans, R. and Zeyrek, M. (2002). *Nucl. Instr. and Meth.* **A485**, 385

would be absorbed in this calorimeter. This particle would decay into two 5 GeV $\gamma$s, and would be assigned an energy value of $2 \times (4.64 + 0.36) = 10$ GeV if the two $\gamma$s would be recognized as two separate showers by the calorimeter. However, if the two showers were so close that the calorimeter would not recognize the event as being caused by two separate particles, it would be assigned an energy of $2 \times 4.64 + 0.36 = 9.64$ GeV, i.e., 3.4% too low. In a non-linear calorimeter, the reconstructed energy thus depends on whether the event in question is caused by one particle or by several particles of which the showers overlap to the point that they are not recognized as such.

Figure 4.2 illustrates this situation for a practical case. Electromagnetic showers caused by different particles of the same total energy give signal distributions centered around different values. This example is useful since it shows that the precision with which the energy of an unknown object can be measured is only determined by the energy resolution (i.e., the width of the curves) if the central value of the signal distributions corresponds to the correct energy (50 GeV), which is not the case here.

Signal *non-linearity* is one of the most worrisome instrumental effects. There are many possible causes. Miscalibration of the signals from the different sections of a longitudinally segmented calorimeter, which is the origin of the effect shown in Fig. 4.2, is a very common cause, discussed in detail in Chap. 9. Hadron calorimeters are usually *intrinsically* non-linear, because the invisible energy (which does *not* contribute to the calorimeter signals) represents an energy dependent fraction of the total. The effects of this are discussed in Chap. 7.

In this chapter, instrumental effects that may lead to signal non-linearity are addressed. We distinguish two types of non-linearity, which have quite different consequences for the calorimeter performance.

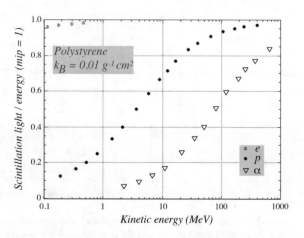

**Fig. 4.3** Suppression of the scintillation light signals for densely ionizing particles because of quenching effects. Results are given for electrons, protons and $\alpha$ particles in polystyrene as a function of the kinetic energy of the particles. From: Wigmans, R. (2018), *J. Progr. Part. Nucl. Phys.* **103**, 109

## 4.2.1   Non-linearity Resulting from Quenching Effects

The first type of non-linearity is caused by *quenching* effects in the signal producing medium. It affects the signals from densely ionizing shower particles. Detector media that are susceptible to this effect include scintillators and noble liquids. The latter are based on ionization charge drifting over a rather long distance to an electrode that converts the collected charge into a measurable signal. If the ionization density in the liquid is large, then the probability for recombination of electrons and ions into atoms along the track increases and the charge collected at the electrode decreases as a result. A similar phenomenon occurs in scintillators, where the effects are well described by Birks' law:

$$\frac{dL}{dx} = S\frac{dE/dx}{1 + k_{\mathrm{B}} \cdot dE/dx} \tag{4.3}$$

where $L$ is the amount of light produced by a particle of energy $E$, $S$ a proportionality constant and $k_{\mathrm{B}}$ a material property known as Birks' constant [2]. This constant is typically of the order of $0.01\,\mathrm{g\,cm^{-2}\,MeV^{-1}}$, whereas the specific ionization ($dE/dx$, also known as the stopping power) of a minimum ionizing particle (mip) is of the order of $1\,\mathrm{MeV\,g^{-1}\,cm^2}$.

Figure 4.3 illustrates the effects of this quenching in polystyrene-based scintillators. It shows that signals from 10 MeV protons are suppressed by $\sim$30% compared to the signals from mips. For 1 MeV protons, the signal reduction is a factor of three. The figure also shows that the effects are much larger for $\alpha$ particles, and negligible for electrons.

Such quenching effects do not play a role in certain other detector media, in particular silicon semiconductors and gases. This may lead to some very troublesome effects in sampling calorimeters that use such detectors as active material. Figure 3.12b shows that 1 MeV protons have a range of about 2 cm in air. The range in

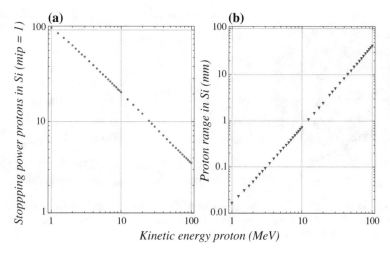

**Fig. 4.4** The stopping power, normalized to the mip value (**a**) and the range (**b**) of densely ionizing protons in silicon. Data from [3]

gases used in proportional wire chambers is very similar. Such a (recoil) proton may be produced in elastic $n - p$ scattering in that gas and deposit its entire 1 MeV energy inside the wire chamber. On the other hand, a mip traversing the same wire chamber typically deposits at least three orders of magnitude less energy in it [3]. This leads to a phenomenon known as the *Texas tower effect*, in which one individual shower particle can mimic a local deposit of an enormous amount of energy. It occurs in calorimeters with a very small sampling fraction and a non-quenching active medium. In Sect. 4.3, several practical examples of this effect are discussed. These examples include calorimeters in which thin silicon sensors are used to produce the signals.

Figure 4.4a shows that the stopping power of silicon is larger than that for mips by a factor ranging from 20 to 100 in the relevant energy range (10–1 MeV) of protons produced in hadron shower development. The amplification factor of a proton signal in such calorimeters is also determined by the range of these particles in the sensors (Fig. 4.4b). The thinner the silicon, the more prone the detector will be to local spikes in the measured energy deposit pattern.

The Texas tower effect does not play a role in calorimeters that use plastic scintillator or liquid-argon as active material, because of the combined effects of signal quenching and the larger thickness of the active layers. And since densely ionizing shower particles represent a more or less energy independent fraction of the (non-$\pi^0$) shower component, the quenching effects do not adversely affect the performance characteristics of such calorimeters. That is different for the second type of non-linearity, which is the result of signal *saturation*.

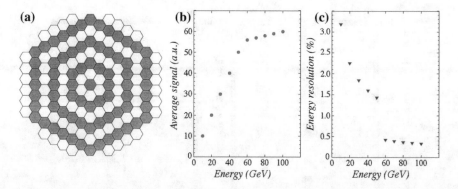

**Fig. 4.5** Saturation effects in one of the towers of the SPACAL calorimeter (**a**). Shown are the average signal (**b**) and the energy resolution (**c**) as a function of energy, measured when a beam of electrons was sent into this tower [4]

## 4.2.2    *Non-linearity Resulting from Saturation*

Saturation is a consequence of the intrinsic limitations of the device that generates the signals. We use data from one of our own experiments to illustrate the effects of signal saturation. The SPACAL calorimeter (Fig. 4.5a) consisted of 155 hexagonal towers. The scintillation light produced in each of these towers was read out by a photomultiplier tube (PMT). Each tower was calibrated by sending a beam of 40 GeV electrons into its geometric center. Typically, 95% of the shower energy was deposited in that tower, the remaining 5% was shared among the six neighbors. The high-voltage settings of the PMT were chosen such that the maximum energy deposited in each tower during the envisaged beam tests would be well within the dynamic range of that tower. For most of the towers (except the central septet), the dynamic range of the PMT signals was chosen to be 60 GeV.

When we did an energy scan with electrons in one of these non-central towers, the results shown in Fig. 4.5b, c were obtained. Up to 60 GeV, the average calorimeter signal increased proportionally with the beam energy, but above 60 GeV a non-linearity became immediately apparent (Fig. 4.5b). The PMT signal in the targeted tower had reached its maximum value, and would from that point onward produce the same value for every event. Any increase measured in the total signal was due to the tails of the shower, which developed in the neighboring towers. A similar trend occurred for the energy resolution (Fig. 4.5c). Beyond 60 GeV, the energy resolution suddenly improved dramatically. Again, this was a result of the fact that the signal in the targeted tower was the same for all events at these higher energies. The energy resolution was thus completely determined by event-to-event fluctuations in the energy deposited in the neighboring towers by the shower tails. These were relatively small because of the small fraction of the total energy deposited in these towers.

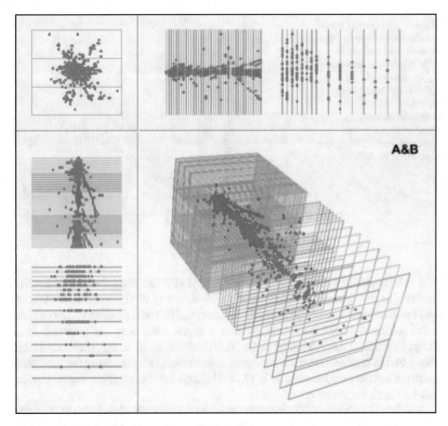

**Fig. 4.6** Event display for a 120 GeV $\pi^-$ showering in the CALICE digital hadron calorimeter. From: Sefkow, F. et al. (2016). *Rev. Mod. Phys.* **88**, 015003

This example illustrates two important consequences of signal saturation:

- Non-linearity of the calorimeter response, i.e., the total calorimeter signal is not proportional to the energy of the detected particle (Fig. 4.5b).
- Overestimated energy resolution. The energy resolution is determined by the combined effects of all fluctuations that may occur in the shower development. Signal saturation leads to the suppression of a certain source of fluctuations, and the actual resolution is thus worse than measured (Fig. 4.5c).

In the above example, these consequences could be easily avoided, by decreasing the high voltage and thus the gain of the PMT that converted the scintillation light into electric signals. This is very different for the devices that produce the signals in so-called *digital* calorimeters.

**Fig. 4.7** The average signal as a function of energy for electrons and hadrons in a large calorimeter based on digital readout. About 500,000 RPC cells were embedded in a structure consisting of tungsten absorber plates. From: Sefkow, F. et al. (2016). *Rev. Mod. Phys.* **88**, 015003

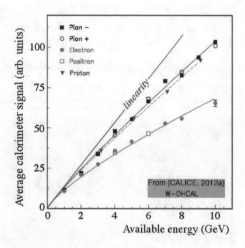

A calorimeter of this type was built by the CALICE Collaboration [5]. The active elements of this detector are resistive plate chambers (RPCs) with in total 500,000 small readout pads (a cross section of $1 \times 1$ cm$^2$ each). These devices, which operate in the saturated avalanche mode, produce a signal when they are traversed by a charged particle. The calorimeter in question produces pretty energy deposit patterns (Fig. 4.6), but is otherwise not a very good calorimeter. This is because the RPC does not make a difference between 1, 3, 17 or 53 charged particles. It is a digital device, with two options: *yes* or *no*.

It is clear that this type of calorimeter exhibits the effects outlined above: response non-linearity (Fig. 4.7) and overestimated energy resolution. These effects are also worse for em showers compared to hadronic ones because of the larger spatial density of shower particles. The described effects could of course be mitigated by reducing the size of the RPC readout pads, but the number of readout channels that would have to be handled in that case might exceed the limit of what is reasonably possible in that respect.

Similar, albeit probably easier to solve, problems are faced when silicon photo-multipliers (SiPM) are used to detect the light signals produced by calorimeters. A SiPM consists of a large number of very tiny pixels, each of which is a photon detector operating in the Geiger mode [6, 7]. At the moment of this writing, SiPMs with up to 40,000 pixels per mm$^2$ are the state of the art [8]. Since these pixels are also digital detectors (yes/no), saturation effects will occur when two or more photons hit the same pixel within the time used to collect the light signals, with the consequences described above. In this case, these consequences might be mitigated by a further reduction of the pixel size and/or a decrease of the intensity of the light signals to which the SiPM is exposed.

## 4.3 The Texas Tower Effect

This term describes a class of phenomena that may be characterized as catastrophic effects caused by one individual shower particle. These effects are directly linked to the absence of signal quenching (Sect. 4.2.1). The origin of the name goes back to the CDF experiment that operated at Fermilab's Tevatron in the period 1985–2011. This experiment used initially in its forward region (i.e., close to the beam pipe) a calorimeter based on isobutane-filled wire chambers as active medium. This calorimeter was built by a group from Texas A&M.

The effect in question is a consequence of the very small sampling fraction of these calorimeters, combined with the fact that the gaseous active medium contained hydrogen. The sampling fraction was $\mathcal{O}(10^{-5})$, which means that the energy deposited by a showering 100 GeV hadron in the active material amounted to ~1 MeV. However, when one of the numerous MeV-type neutrons produced in this process scattered off a hydrogen nucleus, a comparable amount of energy was deposited in the gas layer *by that one recoil proton!*. This led to signal profiles in which anomalously large amounts of energy seemed to be deposited at random positions, sometimes very far away from the shower axis. i.e., the extrapolated trajectory of the showering particle.

In beam tests, where a beam of mono-energetic particles is sent into the calorimeter, such anomalous events are easily recognized, and can be removed from the event samples. However, during the operation of such a calorimeter in an accelerator environment, this is not (always) possible, especially if no additional information about the event in question, e.g., from a tracker system, is available. This difficulty led the CDF Collaboration to the decision to scrap their gas-based forward calorimeter, and replace it with a device that had a sampling fraction that was three orders of magnitude larger [9].

The CMS experiment at CERN's LHC encountered a problem of a very similar nature, although the explanation is somewhat different. When the CMS experiment was designed, much emphasis was placed on excellent performance for the detection of high-energy photons, in view of the envisaged discovery of the Higgs boson through its $H^0 \rightarrow \gamma\gamma$ decay mode. To that end, the Collaboration decided to use PbWO$_4$ crystals for the em section of the calorimeter, since these would provide $\mathcal{O}(1\%)$ energy resolution for the $\gamma$s produced in this process. Since the crystals would have to operate in a strong magnetic field, Avalanche Photo Diodes were chosen to convert the light produced by these crystals into electric signals. Given the available sizes of APDs at that time, and in order to take full advantage of the available (small) light yield, each crystal was equipped with two APDs (Fig. 4.8a). However, in order to save some money, these APDs were ganged together and treated as one device in the data acquisition system.

The hadronic section of the CMS calorimeter consists of a sampling structure, based on brass absorber and plastic scintillator plates as active material. Both sections were developed completely independently, and tested separately in different beam lines at CERN. For the tests of the em section, high-energy electron beams were

**(a)**                          **(b)**                          **(c)**

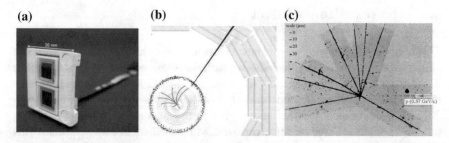

**Fig. 4.8**  Photograph of two CMS APDs (active area $5 \times 5$ mm$^2$) mounted in a capsule (**a**). CMS event display of a $pp$ collision event, showing an isolated ECAL spike (top-right) simulating a 690 GeV transverse energy deposit (**b**). From: Petyt, D.A. (2012). *Nucl. Instr. and Meth.* **A695**, 293. A nuclear reaction induced by a proton with a kinetic energy of 160 MeV in a photographic emulsion (**c**).  Photograph courtesy CERN

used, the hadronic section was exposed to beams of all available types of particles (electrons, pions, kaons, protons, muons). The performance of both sections was documented in detail, and found to be in agreement with expectations [10, 11].

Yet, when the entire calorimeter system was assembled and exposed to high-energy hadrons, an unexpected surprise occurred [12]. In some fraction of the events, anomalously large signals were observed. An example of this phenomenon is shown in Fig. 4.8b. What was going on?

The APDs that convert the light produced in the crystals are also extremely sensitive to ionizing particles. In fact, a mip traversing such an APD may create a signal equivalent to several thousand light quanta [13]. Measurements performed with muons traversing the CMS PbWO$_4$ crystals revealed that a muon that passed through the active layer of an APD generated a signal that was, on average, equivalent to the signal from scintillation photons created by an energy deposit of 160 MeV inside the crystal [14]. The signals produced by densely ionizing charged particles are correspondingly larger.

In Fig. 4.8c, an example is given of interactions that are typical when a high-energy hadron strikes an atomic nucleus. Several densely ionizing nuclear fragments are visible in this picture, with $dE/dx$ values that are up to 100 times larger than that of a mip (see Fig. 4.4a). If such a nuclear interaction would take place in the vicinity of an APD, the nuclear fragments traversing the active detector surface area could produce a very large signal. Given the relationship between the signals from mips and from scintillation photons, such an event could well mimic an energy deposit of 100 GeV or more.

This phenomenon was only discovered when the em and hadronic calorimeter sections were assembled together and exposed to high-energy hadrons. Since the em section corresponds to about one nuclear interaction length, a substantial fraction of hadrons entering the calorimeter start the shower development process in the em section, and therefore the process that generates the "spikes" described above becomes a realistic possibility.

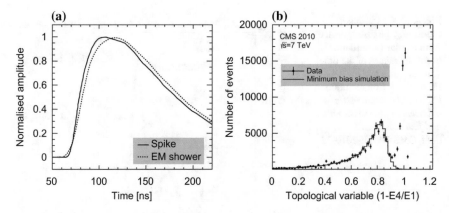

**Fig. 4.9** Distinguishing characteristics of CMS "spike" events. Shown are the average pulse shape (**a**) and the distribution of the so-called *Swiss Cross* topological variable for the highest energy deposit in each event (**b**). See text for details. From: Petyt, D.A. (2012). *Nucl. Instr. and Meth.* **A695**, 293

As stated above, the nuclear reactions have to take place "close to" the sensor surface of the APD in order to produce this effect. The scale on Fig. 4.8c clarifies what "close" means in this context, i.e., within 100 μm or so. This also provides the answer to the question how this problem could have been avoided. Since the two APDs that read out each crystal are separated by several mm, a nuclear interaction of the type discussed here would never affect both APDs, but only one of them. Therefore, if the two APDs had been read out separately, instead of being treated as one detector unit in the data acquisition system, "spike" events would be easily recognized since the signal in only one of the APDs would be anomalously large, while the signal from the other one would still provide the useful information one would like to obtain from the event in question.

However, since the two groups that were responsible for the two calorimeter sections worked in their own individual, separate universes, this problem was only discovered when it was too late to make the corrections needed to avoid it. We include this example here because it illustrates that it is important to realize that a calorimeter built for a given experiment is not a stand-alone device, but is part of an integrated system of detectors. The different components of this detector system may affect the performance of each other and it is important to realize and test this in the earliest possible stage of the experiment.

This phenomenon is very similar, at least in its consequences, to the Texas tower effects in CDF, discussed above. Also here, one low-energy shower particle may cause an event in which an anomalously large amount of energy seems to be deposited in the detector (see Fig. 4.8). Contrary to CDF, CMS has made an effort to deal with the real-life consequences of this phenomenon, and has succeeded in recognizing, and eliminating, affected events to a reasonable extent. This is illustrated in Fig. 4.9, which shows two characteristics that distinguish these "spike" events from the "normal"

**Fig. 4.10** CMS "spike"
events may be distinguished
from regular (scintillation)
events by means of the time
structure. See text for details.
From: CMS Collaboration
(2010). *Electromagnetic
calorimeter commissioning
and first results with 7 TeV
data*, CMS-NOTE-2010-12

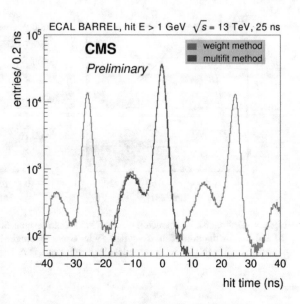

ones. One effective method is a cut on the so-called *Swiss Cross* variable, in which
the signal in each tower of the em calorimeter is compared to the sum of the signals
in the four neighboring towers. Monte Carlo simulations showed that a cut of events
in which $E4/E1 < 0.05$ is an effective tool for eliminating the "spike" events [12].

Also the time structure of the spike events is a differentiating characteristic. Since
the signals in this case are caused by shower particles traversing the APDs, they are
faster than the signals based on detection of the scintillation light generated by the
shower particles in the crystals. The latter are delayed because the molecules excited
in the scintillation process take some time to decay ($\sim$10 ns). Figure 4.10 shows the
time distribution of the hits in the ECAL barrel with a reconstructed energy above
1 GeV, for an 80 ns time slice of the LHC operations. The 25 ns bunch structure is
clearly visible. The events caused by the scintillation light are characterized by the
peaks at $-25$ ns, 0 ns and $+25$ ns in this plot. These peaks are preceded by smaller
peaks that occur about 10 ns earlier. These are the spike events, and the figure shows
that these represent about 3% of the total. It should be emphasized that this plot
concerns the time characteristics of the events that *survived* the Swiss Cross cuts.

Based on the understanding of the underlying cause of this phenomenon, it is
expected that the frequency of these events will increase both with the luminosity
and with the center-of-mass energy of the $pp$ collisions in the LHC. The reason
why this example is included in this context is that it is an inherent feature of the
CMS calorimeter system. Even though attempts to deal with the problem may seem
successful, it is good to keep in mind that any data selection based on the calorimeter
information alone may lead to a biased event sample. Also, it is inevitable that some
fraction of the events in which the process that causes a "spike" occurs will *not* be
eliminated by the cuts devised to deal with the problem.

Given the origin of the problems described in this subsection, other examples of catastrophic effects caused by a single shower particle may be expected in the High-Granularity calorimeter that is scheduled to replace the radiation damaged endcaps of the CMS calorimeter [15] in 2025. For example, in the hadronic section of this instrument, 5 cm thick brass plates will be interleaved with 200 $\mu$m thick silicon sensors that act as active material. The sampling fraction for mips is thus $6 \cdot 10^{-4}$ in this (FH) section, and even less for showers. A nuclear interaction such as the one shown in Fig. 4.8c may easily deposit 30 MeV in the silicon sensor if the event takes place in the vicinity of the boundary between the active and passive material. Since the silicon signals do not saturate for densely ionizing particles, this event will be interpreted as a 50 GeV highly localized energy deposit. The nuclear reactions responsible for this phenomenon are typically initiated by spallation neutrons with kinetic energies of a few hundred MeV. Such neutrons may travel tens of centimeters away from the shower axis before initiating the reaction that gives rise to the large signal. They will also be prolifically produced in the absorption of soft hadrons that constitute the "pileup" component of the events CMS is looking for, and one should thus expect very large signals at (multiple) random locations in the calorimeter, for essentially every bunch crossing. Just like the Texas Tower effect, the described phenomena are a typical consequence of the development of hadron showers in a calorimeter with a small sampling fraction and a non-saturating active medium, and therefore play no role in calorimeters based on plastic-scintillator or liquid-argon readout.

## 4.4  Signals from External Sources

A third class of problems encountered in operating calorimeters is caused by particles that are not supposed to contribute to the signals, since they have nothing to do with the event that is being studied. Also in this case, the CMS calorimeter system has provided a useful example. The very-forward section of the calorimeter system ($3 < \eta < 5$) is based on quartz fibers embedded in an iron absorber structure (Fig. 4.11). High-purity quartz, chosen because of its radiation hardness, provides signals consisting of Čerenkov light generated in it by charged relativistic shower particles, and the fiber structure makes it possible to transport this light to the outside world, where it is converted into electric signals by means of PMTs.

In order to save money, the packing fraction of the quartz fibers was reduced to the point where one detected photoelectron corresponds to 5 GeV deposited energy. In other words, a 100 GeV particle absorbed in this calorimeter provides a signal of about 20 photoelectrons. However, relativistic charged particles that traverse the glass windows of the PMTs generate Čerenkov light in this glass that leads to signals of the same (or larger) amplitude. There is no shortage of such particles during the LHC operations. The proton beams are surrounded by an intense halo of high-energy muons, which are barely affected by the material constituting the CMS detectors. When these muons traverse a PMT that is looking for light signals

**Fig. 4.11** Three modules of the CMS forward calorimeter, which consists of $2 \times 12$ such wedges. The quartz fibers that serve as the active material in this sampling calorimeter are bunched towards a readout box, where PMTs convert the Čerenkov light signals into electric pulses. Photograph courtesy CERN

created by showers inside the calorimeter, they generate signals that in practice may make the information provided by the very-forward calorimeters rather meaningless for the physics analyses, and most definitely renders these calorimeters useless for triggering purposes.

The solution envisaged for this problem is to replace the readout by sensors that are less subject to this effect [16]. In a first stage, the replacement sensors will be multi-anode PMTs with thinner glass windows. The idea is that light produced in the calorimeter will cause similar signals in all anodes of the PMT, while the signal from one single anode will dominate when (localized) light produced by a particle traversing the glass window of a PMT is detected. In addition, algorithms intended to distinguish the stray muon signals from those generated by particles absorbed in the calorimeter are being developed. To that end, the time structure of the events is being used, just as was done to handle the spike events discussed in Sect. 4.3 (Fig. 4.10). In a later stage, CMS plans to replace the entire readout by a system based on silicon photomultipliers.

# References

1. Adloff, C., et al.: Nucl. Instrum. Methods **A608**, 372 (2009)
2. Birks, J.B.: The Theory and Practice of Scintillation Counting. Pergamon, Oxford (1964)
3. Berger, M.J.: Stopping Power and Range Tables for Electrons, Protons and Helium Ions, report NISTIR 4999 (1993). http://physics.nist.gov/Star
4. Wigmans, R.: Calorimetry—Energy Measurement in Particle Physics, 2nd edn. International Series of Monographs on Physics, vol. 168. Oxford University Press, Oxford (2017)
5. Sefkow, F., et al.: Rev. Mod. Phys. **88**, 015003 (2016)
6. Renker, D., Lorenz, E.: J. Instrum. **4**, P04004 (2009). and references therein
7. Garutti, E.: J. Instrum. **6**, C10003 (2011)
8. Acerbi, F., et al.: J. Sel. Top. Quantum Electron. **24**(2), 3800608 (2018)
9. Albrow, M., et al.: Nucl. Instrum. Methods **A480**, 524 (2002)
10. Adzic, P., et al.: JINST **2**, P04004 (2007)
11. Abdullin, S., et al.: Eur. Phys. J. C **55**, 159 (2008)
12. Petyt, D.A.: Nucl. Instrum. Methods **A695**, 293 (2012)
13. Hauger, J.A., et al.: Nucl. Instrum. Methods **A337**, 362 (1994)
14. Alexeev, G., et al.: Nucl. Instrum. Methods **A385**, 425 (1997)
15. CMS Collaboration: Technical proposal for the phase-II upgrade of the Compact Muon Solenoid, CERN-LHCC-2015-10 (2015)
16. Gülmez, E.: (2017). arXiv:1705.00362 [physics.ins-det]

# Part II
# Shower Aspects Important for Calorimetry

Part II
Discover Aspects Important
for Calf rearing

# Chapter 5
# Containment and Profiles

## 5.1 Introduction

When designing a calorimeter system for a certain experiment, it is important that the detector be sufficiently large to contain the showers of interest at an adequate level. Shower particles escaping from the detector represent a source of fluctuations that may affect the precision of the measurements. Another important design consideration is the segmentation of the calorimeter into separate independent sections, which record part of the signals generated by the developing shower. This segmentation is, for example, intended to

- recognize and separate particles that develop showers in each other's vicinity,
- identify the showering particle, and
- measure the direction of the showering particle

The energy deposit profiles of the showering particles are of course a crucial consideration for the design decisions in that respect.

In this chapter, we present information on the absorber size needed to contain the showers, *on average*, at a certain level, e.g., 95%, and on the shower profiles. It turns out that the measured shower profiles are not necessarily equivalent to the energy deposit profiles, but may depend on the signal generation mechanism. It should also be emphasized that the average containment level in itself is no indication of the effects of shower leakage on the energy resolution and other aspects of the calorimeter quality. These effects are determined by *event-to-event fluctuations* about this average. It turns out that 5% longitudinal shower leakage has much larger effects in that sense than 5% transverse shower leakage. This is further discussed in Sect. 6.6.

---

The original version of this chapter was revised: Figures 5.13 and 5.15 have been corrected. The correction to this chapter can be found at https://doi.org/10.1007/978-3-030-23653-3_14.

© Springer Nature Switzerland AG 2019                                         93
M. Livan and R. Wigmans, *Calorimetry for Collider Physics, an Introduction,*
UNITEXT for Physics, https://doi.org/10.1007/978-3-030-23653-3_5

## 5.2   Electromagnetic Showers

### 5.2.1   Shower Containment

Electromagnetic calorimeters are specifically intended for the detection of energetic electrons and $\gamma$s, but produce usually also signals when traversed by other types of particles. They are used over a very wide energy range, from the semiconductor crystals that measure $X$-rays down to a few keV to shower counters such as AGILE, PAMELA and FERMI, which orbit the Earth on satellites in search for electrons, positrons and $\gamma$s with energies $>10$ TeV [1].

Because of the peculiarities of em shower development, these calorimeters don't need to be very deep, especially when high-$Z$ absorber material is used. For example, when 100 GeV electrons enter a block of lead, $\sim$90% of their energy is deposited in only 4 kg of material.

Good energy resolution can only be achieved when the shower is, on average, sufficiently contained. Figure 5.1a shows the required depth of the calorimeter, as a function of the electron energy, needed for 99% longitudinal containment. The figure shows this depth requirement, needed for energy resolutions of 1% (cf. Fig. 6.10), for calorimeters using Pb, Sn, Cu and Al as absorber material. The fact that the four curves are not identical indicates that the radiation length ($X_0$), which is commonly used to describe the longitudinal development of em showers, is not a perfect scaling variable (cf. Fig. 3.10). It should also be noted that $\gamma$-induced showers require about one radiation length more material in order to be contained at a certain level than do electron showers of the same energy [3].

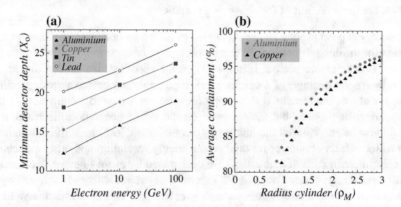

**Fig. 5.1** Size requirements for electromagnetic shower containment. The depth of a calorimeter needed to contain electron showers, on average, at the 99% level, as a function of the electron energy. Results are given for four different absorber media (**a**). Average lateral containment of electron-induced showers in a copper and an aluminium based calorimeter, as a function of the radius of an infinitely deep cylinder around the shower axis (**b**) [2]

## 5.2.2 Shower Profiles

Longitudinal em shower profiles were discussed in Sect. 3.1. The *lateral spread* of em showers is caused by two effects:

1. Electrons and positrons move away from the shower axis because of multiple Coulomb scattering.
2. Photons and electrons produced in more isotropic processes (Compton scattering, photoelectric effect) move away from the shower axis. Also, bremsstrahlung photons emitted by electrons that travel at a considerable angle with respect to the shower axis may contribute to this effect.

The first process dominates in the early stages of the shower development, while the second process is predominant beyond the shower maximum, particularly in high-$Z$ absorber media.

When the shower development in the plane perpendicular to the direction of the incoming particle is discussed, the way data are presented frequently leads to confusion. There are two different ways of presenting such data:

1. The energy density, i.e., the amount of energy per unit volume, is shown as a function of the distance between that unit volume and the shower axis. We will refer to energy distributions and shower profiles of this type as *lateral* energy distributions and *lateral* shower profiles.
2. The energy contained in a radial slice of a certain thickness is shown as a function of the distance between that slice and the shower axis. We will refer to energy distributions and shower profiles of this type as *radial* energy distributions and *radial* shower profiles.

In addition, we will use the term *transverse*, as well as radial or lateral, to describe general, non-quantitative features of shower development in the plane perpendicular to the direction of the incoming particle.

Of course, distributions/profiles of type #1 are considerably narrower than those of type #2. In practice, both terms are being used for both types of distributions/profiles, and it is sometimes not clear which of these two is shown in presentations. In this book, we use the terminology defined above in a consistent manner.

Both types of distributions may either concern a certain longitudinal slice of the absorbing structure, or the entire absorbing structure. In the latter case, the lateral or radial profiles are said to be *integrated over the full depth*.

Figure 5.2 shows the radial distributions of the energy deposited by 10 GeV electron showers developing in copper, at various depths. The two mentioned components can be clearly distinguished. Both show an exponential behavior (note the logarithmic ordinate in Fig. 5.2), with characteristic slopes of $\sim 3$ mm ($\sim 0.2\rho_M$) and $\sim 25$ mm ($\sim 1.5\rho_M$), respectively. The radial shower profile shows a pronounced central core (the first component), surrounded by a *halo* (the second component). The central core disappears beyond the shower maximum.

**Fig. 5.2**  The radial distributions of the energy deposited by 10 GeV electron showers in copper, at various depths. Results of EGS4 calculations [4]

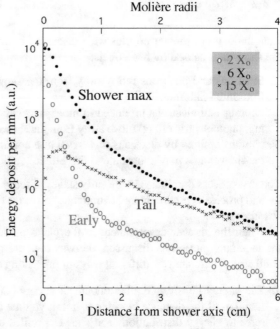

**Fig. 5.3**  Radial energy deposit profiles for 10 GeV electrons showering in aluminium, iron and lead. Results of EGS4 calculations [4]

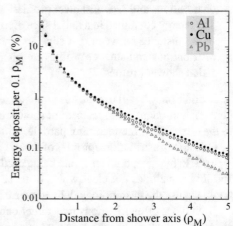

This radial profile, integrated over the total shower depth, is shown in Fig. 5.3, together with the equivalent profiles for 10 GeV electron showers developing in lead and aluminium. The distance from the shower axis (plotted horizontally) is expressed in Molière units. Scaling with $\rho_M$ would imply that these three profiles are identical. The figure shows that this is approximately true, but that there are also some clear differences between these three profiles. And just as in the case of the *longitudinal*

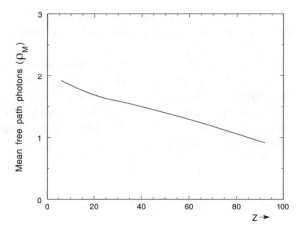

**Fig. 5.4** Mean free path, in units of the Molière radius, for photons with energies 1–3 MeV, as a function of the $Z$ value of the absorber material [4]

shower development, these differences can be understood from phenomena taking place at the 1 MeV level, for which the Molière radius has no meaning.

The most striking difference between these radial profiles concerns the slope of the halo. This slope is considerably steeper for lead ($\sim 1.1\rho_M$) than for the low-$Z$ materials, aluminium and copper ($\sim 1.6\rho_M$). As we saw above, this halo component of the radial em shower profiles is predominantly due to soft electrons, with energies well below the critical energy, which are produced in Compton scattering and photoelectric absorption. The slope of the halo is thus directly related to the mean free path of the photons causing these processes.

The mean free path of photons in the 1 MeV energy range is typically considerably different from that of GeV-type photons, which convert on average after $\frac{9}{7}X_0$. This mean free path ($\langle l \rangle$, expressed in g cm$^{-2}$) can be calculated from the total cross section ($\sigma$), to which it is related as

$$\sigma(E) = \frac{A}{N_A \langle l \rangle} \tag{5.1}$$

For example, in lead, the asymptotic cross section for photon interactions amounts to 42 barns (Eq. 3.1), whereas the cross section drops to 15.6 b for $E_\gamma = 2$ MeV [5]. Therefore, the mean free path of $\gamma$s of a few MeV equals $42/15.6 \times 9/7 = 3.5$ $X_0$. That is why radioactive $^{60}$Co sources, which emit $\gamma$s of about 1.3 MeV, require substantial lead shielding. These $\gamma$s are among the most penetrating ones available, and therefore the most difficult ones to shield.

We have used Eq. 5.1 and the Tables from [5] to calculate the mean free paths for $\gamma$s in the energy range from 1–3 MeV for a variety of absorber materials with different $Z$ values and have converted these into units of the Molière radius. The results are given in Fig. 5.4. It turns out that this distribution gradually decreases with increasing $Z$. In the Al–Cu region, the mean free path of these photons amounts to $1.6 - 1.8\rho_M$, while for lead it has dropped to $\sim 1.0\rho_M$. These values are in good agreement with the observed slopes of the halo in Fig. 5.3.

Another effect that may contribute to the less steep halo slopes in low-$Z$ materials is that the absorption of low-energy bremsstrahlung photons requires, on average, more steps than in high-$Z$ materials. In Compton scattering, a photon of lower energy is produced and this photon needs to be absorbed in a subsequent process. In the energy range from 0.2–0.5 MeV, Compton scattering is the most likely process in aluminium and iron (see Fig. 2.5), while in lead, photons in this energy range are predominantly absorbed in a single-step process, photoelectron production.

### 5.2.3   Experimental Data

All the results shown so far in this subsection were obtained with (EGS4) Monte Carlo simulations. However, available experimental results confirm all the trends mentioned above. Detailed measurements on em shower profiles were carried out by Bathow and coworkers [6]. They studied the three-dimensional profiles induced by 6 GeV electrons in aluminium, copper and lead. Their experimental data were provided by arrays of tiny (a few mm$^3$) silver-phosphate dosimeter glasses which were installed at various depths inside the absorber blocks. In these glasses, luminescence centers were formed by ionizing radiation. Afterwards, the accumulated dose could be determined, with a relative precision of about 5%, by exposing the irradiated dosimeters to ultraviolet light.

The absorber blocks with the built-in dosimeters were irradiated with a 6 GeV electron beam provided by the DESY synchrotron, at a rate of $5 \cdot 10^{10}$ particles per second. The dosimeter matrix thus accumulated a three-dimensional image, corresponding to the average shower profile of a very large number of 6 GeV electrons, that was "frozen" into the block and that could be analyzed offline.

Figure 5.5 shows some results of these experiments. In Fig. 5.5a, the longitudinal shower profiles in aluminium, copper and lead are given, with the depth expressed in units of $X_0$. These curves show the same characteristics as the simulated ones in Fig. 3.10: approximate scaling with $X_0$, but as $Z$ increases, the shower maximum shifts to greater depth and the slope beyond the shower maximum becomes less steep.

In Fig. 5.5b, the lateral profile in lead is shown, at three different depths. As in the simulated curves (Fig. 5.2), there are clearly two components visible, the steepest of which disappears beyond the shower maximum. Figure 5.5c shows the energy deposition, integrated over the full shower depth and plotted as a function of the distance from the shower axis, in units of $\rho_M$. The latter curves also exhibit the two-component structure, and the long-distance component is steeper for lead than for copper, as in the EGS4 simulations (Fig. 5.3).

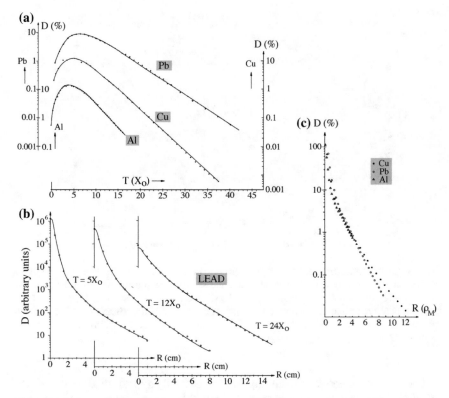

**Fig. 5.5** Experimental results on the shower profiles of 6 GeV electrons in aluminium, copper and lead. Shown are the longitudinal profiles in these three materials (**a**), the lateral profiles in lead, measured at 3 different depths (**b**), and the energy deposition integrated over depth as a function of the distance to the shower axis, for different absorber materials (**c**). Data from [6]

## 5.2.4 Dependence on the Detection Mechanism

It is commonly assumed that the radial energy deposit profile of em showers scales with the Molière radius. For this reason, experiments typically choose the granularity of their em calorimeter in terms of that parameter. For example, if a cell size with an effective radius of $1\rho_M$ is used, em showers deposit typically $\sim$80% of their energy in one cell, if the particle enters that cell in its central region (Fig. 5.1b). In order to increase that percentage to 90%, the effective radius of the cell has to be doubled, i.e., the number of cells is reduced by a factor of four. Recent measurements with a calorimeter that used a much finer granularity showed that the energy deposit profile is very strongly concentrated near the shower axis [7].

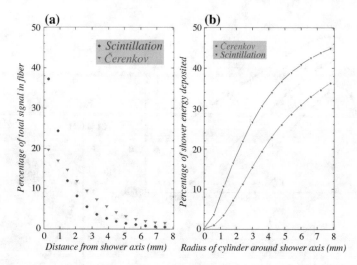

**Fig. 5.6** Lateral profiles of electromagnetic showers in the brass-fiber dual-readout SiPM calorimeter, measured separately with the Čerenkov and the scintillation signals (**a**). The fraction of the shower energy deposited in a cylinder around the shower axis as a function of the radius of that cylinder, measured separately with the Čerenkov and the scintillation signals (**b**). One mm corresponds to $0.031 \rho_M$ in this calorimeter. From: Antonello, M. et al. (2018). *Nucl. Instr. and Meth.* **A899**, 52

For example, when the granularity was increased by a factor of 20 (reducing the effective radius of a detector cell from $1 \rho_M$ to $0.22 \rho_M$), one cell contained ~45% of the shower energy, and if the granularity was increased by another factor of 30 (to cells with a radius of $0.04 \rho_M$), ~10% of the total shower energy was still deposited in one cell. Figure 5.6 shows the profiles measured with this detector, a dual-readout calorimeter (see Sect. 8.3) with fibers read out by a silicon photomultiplier array. The signals from each individual fiber were sensed by a 1 mm$^2$ SiPM [7].

Figure 5.6a shows a remarkable difference between the profiles measured by the two types of fibers that constitute the active material of this calorimeter. The Čerenkov light is much less concentrated near the shower axis than the scintillation light. This is a consequence of the fact that the early, extremely collimated component of the developing shower does not contribute to the Čerenkov signals, since the Čerenkov light falls outside the numerical aperture of the fibers. This phenomenon has interesting consequences for the detection of muons with a device of this type, since a comparison between the two signals makes it possible to distinguish the ionization and radiative components of the energy loss by muons traversing this calorimeter (see Sect. 11.4.1 [8]).

## 5.3  Hadronic Showers

### 5.3.1  Longitudinal Shower Containment

When an experiment is designed, one of the most important decisions concerns the total thickness of the calorimeter. Especially in a $4\pi$ geometry, this decision has serious consequences for the cost of the experiment. Let us, as an example, consider a spherical calorimeter that surrounds the interaction vertex, starting at a distance of 1 m. Let us also assume that the effective nuclear interaction length of this calorimeter is 20 cm. If we want to make this calorimeter $7\lambda_{int}$ thick, then its total volume amounts to $\frac{4}{3}\pi(2.4^3 - 1) = 53.7$ m$^3$. Should we want to add one extra interaction length, this volume would increase by 29% (69.4 m$^3$), while the surface area of detectors installed outside the calorimeter would increase by 17%. And since the cost of many detectors is more or less proportional to the instrumented mass, a decision to go from $7\lambda_{int}$ to $8\lambda_{int}$ would have major financial implications.

In this subsection, we discuss the absorber thickness needed to contain hadron showers, *on average*, at a certain level. We re-emphasize what was said in Sect. 5.1, namely that the effects of shower leakage on the quality of the calorimeter data is determined by *event-to-event fluctuations* about this average, and not by the average shower containment itself. Both for em and for hadron showers, these fluctuations are much larger for longitudinal leakage than for lateral shower leakage, at a given level of shower containment. The reasons for this are discussed in Sect. 6.6.

Because of the practical implications mentioned above, there is plenty of experimental information about hadronic shower containment. Representative results, obtained by WA1 for hadron absorption in iron [9], are shown in Fig. 5.7. The average shower fraction contained in the absorber material is shown as a function of the absorber thickness, for showering pions with energies ranging from 10 GeV to 138 GeV. The absorber thickness needed to contain 95% of the shower energy ranges from ~$3\lambda_{int}$ at 10 GeV to more than $6\lambda_{int}$ at 138 GeV. For 99% containment, the absorber thickness has to be at least $5\lambda_{int}$ deep for 10 GeV pions and ~$9\lambda_{int}$ for 138 GeV ones.

**Fig. 5.7**  Average energy fraction contained in a block of matter with infinite transverse dimensions, as a function of the thickness of this absorber, expressed in nuclear interaction lengths. Shown are results for showers induced by pions of various energies in iron absorber [4]. Experimental data from [9]

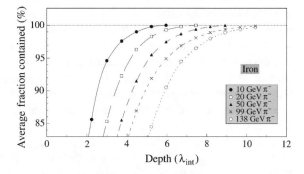

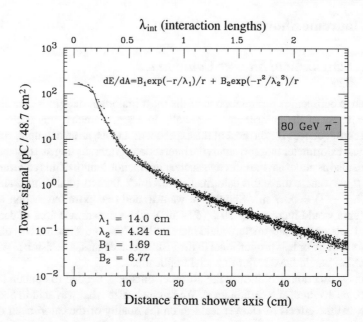

**Fig. 5.8** Average lateral profile of the energy deposited by 80 GeV $\pi^-$ showering in the SPACAL detector. The collected light per unit volume is plotted as a function of the radial distance to the impact point. From: Acosta, D. et al. (1992). *Nucl. Instr. and Meth.* **A316**, 184

## 5.3.2  Shower Profiles

In Sect. 3.2, longitudinal hadron shower profiles, as well as differences with em ones are discussed. Hadron showers do not only start to develop until reaching a much greater depth inside the absorber material, they are also considerably *broader* than electromagnetic showers. The lateral shower profiles exhibit in most materials a narrow core, surrounded by a halo. A representative shower profile, integrated over the full depth of the absorber, is shown in Fig. 5.8. This profile was measured with the SPACAL detector [10].

The narrow core represents the electromagnetic shower component, caused by $\pi^0$s produced in the shower development. The halo, which has an exponentially decreasing intensity, is caused by the non-electromagnetic shower component. A detailed comparison of lateral profiles measured with the SPACAL detector showed that the radius of the cylinder around the shower axis needed to contain 80 GeV $\pi^-$ showers at the 95% level is about 32 cm ($1.5\lambda_{int}$), nine times larger than the 3.5 cm ($1.8\ \rho_M$) radius for containing 80 GeV em showers at the same level [10].

Longitudinally, the difference in the amounts of material needed for containing these two types of showers at a certain level is very similar (i.e., a factor of about nine). Measurements showed that the average energy fraction leaking out at the back of the $9.6\lambda_{int}$ deep SPACAL detector amounted to $\sim$0.3%, for showers induced by 80 GeV $\pi^-$ [11]. An average longitudinal containment of 99.7% for 80 GeV electron

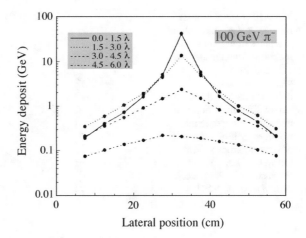

**Fig. 5.9** Lateral profiles for pion-induced showers, measured at different depths, with the ZEUS calorimeter [4]. Experimental data from [12]

showers requires 30 $X_0$ worth of lead, and $(9.6 \times 170)/(30 \times 5.6) \approx 9$ times more for 80 GeV pions.

Hadron showers are thus larger in *all* spatial dimensions. In the example of SPACAL, it took $9 \times 9 \times 9 \approx 700$ times as much material to contain 80 GeV pion showers at the same level as electron showers of the same energy. This offers multiple possibilities to identify particles on the basis of their shower profile characteristics, both the longitudinal and the lateral ones.

Next, we take a look at the *differential* lateral shower profiles, i.e., lateral shower profiles measured at different depths inside the absorber. Not surprisingly, the width of these profiles gradually increases with depth, as shown in Fig. 5.9. This figure shows the results of measurements performed by the ZEUS Collaboration on the energy deposit profiles of 100 GeV pions showering in their uranium/plastic-scintillator calorimeter [12]. The electromagnetic shower core is very prominently present in the initial stages of the shower development, but has largely disappeared beyond a depth of $4.5\lambda_{int}$.

Interesting information about hadronic shower profiles was obtained by Leroy and coworkers, who exposed stacks of large numbers of thin metal plates to beams of high-energy hadrons and analyzed the distributions of radioactive nuclides produced in this process [13]. These measurements revealed detailed features of the lateral profiles. Figure 5.10 shows the lateral distributions of several radioactive nuclides, measured at a depth of $4\lambda_{int}$ inside a block of uranium. The distributions exhibit considerable differences. The distribution of the $^{239}$Np nuclei is much broader than that of $^{99}$Mo, which in turn has a larger width than the $^{237}$U distribution. These distributions are different because the mechanisms through which these nuclides were produced are different.

Neptunium-239 is the decay product of $^{239}$U, which is produced when $^{238}$U, the absorber material in these experiments, captures a neutron. This is by far most likely to happen for thermal neutrons. In other words, the $^{239}$Np distribution is a measure

**Fig. 5.10** Lateral profiles for 300 GeV $\pi^-$ interactions in a block of uranium, measured from the induced radioactivity at a depth of $4\lambda_{int}$ inside the block. The ordinate indicates the decay rate of different radioactive nuclides, produced in nuclear reactions by different types of shower particles. From: Leroy, C., Sirois, Y. and Wigmans, R. (1986). *Nucl. Instr. and Meth.* **A252**, 4

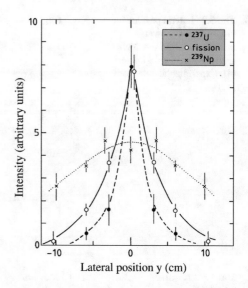

for the spatial density of thermal neutrons at a depth of $4\lambda_{int}$ inside the block of depleted uranium.

The nuclide $^{99}$Mo is a fission product of uranium. The threshold for neutron-induced $^{238}$U fission is about 1.5 MeV. Therefore, the distribution of this radioactive nuclide is a measure for the spatial distribution of the non-thermalized, MeV-type neutrons. Obviously, these neutrons have traveled a much smaller distance from their point of origin and are therefore much more concentrated around the shower axis than the thermal neutrons.

Finally, $^{237}$U is most likely produced through the reaction $^{238}$U $(\gamma, n)$ $^{237}$U. The cross section for this process reaches a maximum value of $\sim$0.35 b, at a photon energy of about 11 MeV [14]. Gammas of this energy are abundantly produced in the em showers generated by $\pi^0$s. Therefore, one expects to find $^{237}$U concentrated in the narrow shower core (Fig. 5.8), close to the shower axis.

### 5.3.3   Lateral Shower Containment

Transverse shower containment results are shown in Fig. 5.11, where the energy fraction contained in a cylinder around the shower axis is plotted as a function of the radius of this cylinder, for pions of several energies showering in lead absorber [10]. Unlike for em showers, the results do depend in this case on the energy of the showering particle, in a way that at first sight seems counter-intuitive: the higher the energy of the incoming particle, the narrower the cylinder needed to contain the shower. For example, to contain 10 GeV pions at the 95% level, a cylinder with a radius of $\sim$1.7$\lambda_{int}$ is needed, while 1.4$\lambda_{int}$ is enough for 150 GeV pions.

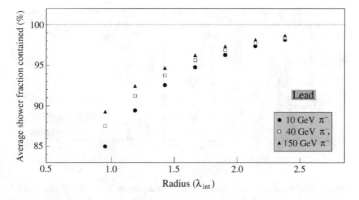

**Fig. 5.11** Average energy fraction contained in an infinitely long cylinder of absorber material, as a function of the radius of this cylinder (expressed in nuclear interaction lengths), for pions of different energy showering in lead absorber. From: Acosta, D. et al. (1992). *Nucl. Instr. and Meth.* **A316**, 184

This energy dependence is a direct consequence of the energy dependence of the average em fraction of the hadron showers ($\langle f_{em} \rangle$), discussed in detail in Sect. 7.2. The average energy fraction carried by the em shower component increases with energy and since this component is concentrated in a narrow core around the shower axis, the energy fraction contained in a cylinder with a given radius increases with energy as well.

## 5.3.4 Dependence on Hadron Type

A different calorimeter response to the em and non-em components of hadron showers also leads to differences in the response functions for different types of hadrons. The absorption of different types of hadrons in a calorimeter may differ in very fundamental ways, as a result of applicable conservation rules. For example, in interactions induced by a proton or neutron, conservation of baryon number has important consequences. The same is true for strangeness conservation in the absorption of kaons. This has implications for the way in which the shower develops. For example, in the first interaction of a proton, the leading particle has to be a baryon. This precludes the production of an energetic $\pi^0$ that carries away most of the proton's energy. Similar considerations apply in the absorption of strange particles. On the other hand, in pion-induced showers it is not at all uncommon that most of the energy carried by the incoming particle is transferred to a $\pi^0$. The resulting shower is in that case almost completely electromagnetic. This phenomenon is the reason for an asymmetric signal distribution. It also leads to a smaller value of $\langle f_{em} \rangle$ in baryon induced showers, compared to pion induced ones.

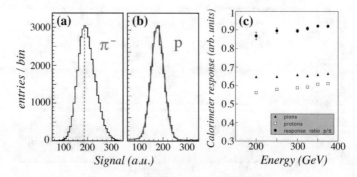

**Fig. 5.12** Signal distributions for 300 GeV pions (**a**) and protons (**b**) in the CMS forward calorimeter. Average signals per GeV for protons and pions as well as the ratio of these response values
in this detector, as a function of energy (**c**). From: Akchurin, N. et al. (1998). *Nucl. Instr. and
Meth.* **A408**, 380

Experimental studies have confirmed these effects [15, 16]. Figure 5.12 shows
the signal distributions measured for 300 GeV pions (*a*) and protons (*b*), respectively. The signal distribution for protons is much more symmetric, as indicated by
the Gaussian fit. This is because the em component of proton-induced showers is
typically populated by $\pi^0$s that share the energy contained in this component more
evenly than in pion-induced showers. The figure also shows that the rms width of
the proton signal distribution is significantly smaller (by $\sim$20%) than for the pions.
Figure 5.12c shows that the average signal per GeV deposited energy is smaller for
the protons than for the pions, by about 10%. This is also a consequence of the
limitations on $\pi^0$ production that affect the proton signals. So while the response to
protons is smaller in this calorimeter, the energy resolution is better. Similar effects
are expected to play a role for the detection of kaons, where $\pi^0$ production is limited
as a result of strangeness conservation in the shower development.

The ATLAS Collaboration used the longitudinal profile data to determine the
shower leakage expected in their calorimeter system. Since the effective calorimeter
thickness increases with the pseudorapidity ($\eta$), the leakage decreases with the angle
between the particle trajectory and the beam line. Figure 5.13 shows the average
energy leakage for 100 GeV pions and protons as a function of $\eta$. The corresponding effective calorimeter thickness is plotted on the top axis. The smaller leakage
observed for protons is a reflection of the fact that the shower profile for these particles is a bit shorter because of the smaller interaction length, combined with leading
particle effects [16].

**Fig. 5.13**  The average
shower leakage for 100 GeV
pions and protons in the
ATLAS calorimeter system,
as a function of
pseudorapidity [4].
Experimental data from[16]

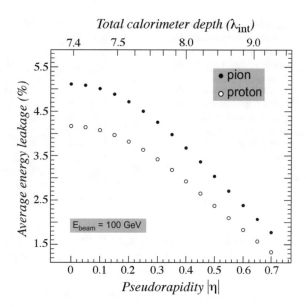

### 5.3.5   Dependence on the Detection Mechanism

Only relativistic, charged shower particles contribute to the signals from calorimeters based on Čerenkov light. Their velocity should exceed the Čerenkov threshold: $v > c/n$, where $n$ represents the refraction index of the medium in which the particles travel. Media that are frequently used in Čerenkov calorimeters include water ($n = 1.33$), quartz ($n = 1.46$) and various types of lead-glass ($n = 1.5$–$1.75$). Shower particles that may contribute to the signals include

- Electrons and positrons. For $n$ values of $\sim 1.4$, these particles emit Čerenkov light when their kinetic energy exceeds 200 keV.
- Charged pions. In typical calorimeter media, these short-lived particles emit Čerenkov light when their total energy is larger than about 190 MeV.
- Protons. These need to carry a kinetic of at least 400 MeV in order to generate Čerenkov light in such calorimeters.

The signals from Čerenkov calorimeters depend completely on the extent to which Čerenkov-capable particles are produced in the shower development. As we have seen before, hadronic shower development involves various processes, each with very different rates of Čerenkov-capable particle production:

- $\pi^0$s produced in hadronic shower development give rise to em showers. Most of the energy of these $\pi^0$s is deposited through electrons and positrons above the Čerenkov threshold (kinetic energy larger than 200 keV).
- Of the energy carried by the non-electromagnetic shower component, on average $\sim 20\%$ is deposited by charged pions.

**Fig. 5.14** A comparison of
the transverse characteristics
of 80 GeV $\pi^-$ showers
measured with a scintillation
calorimeter [10] and with a
Čerenkov calorimeter [17].
Shown is the fraction of the
signal recorded outside a
cylinder with radius $R$
around the shower axis, as a
function of $R$ [4]

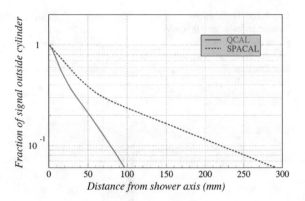

- The rest of the non-electromagnetic energy is deposited by protons and neutrons, almost all of which are non-relativistic, and through release of nuclear binding energy, which leaves no directly measurable signal.

Since electrons and positrons produced in the showers dominate the signals from this calorimeter, hadron showers thus register predominantly through their electromagnetic shower core. This has several important consequences.

One of these consequences concerns the (three-dimensional) hadronic shower profiles. The instrumented volume needed to contain the Čerenkov-capable shower component is substantially smaller than that required for full containment of the entire hadron shower. This is true both in depth (longitudinal containment) and in the transverse plane, since the shower tails in all directions are primarily composed of non-relativistic particles (soft nucleons).

For this reason, hadron showers in Čerenkov calorimeters appear to be considerably narrower than in other types of calorimeters. This may be an important advantage when particle densities are very high (for example, in the high-$\eta$ region of LHC experiments), since it reduces shower overlap.

This point is illustrated in Fig. 5.14, which shows transverse characteristics of showers, initiated by 80 GeV $\pi^-$ mesons in the Quartz Fiber Calorimeter (Čerenkov light, copper absorber) and in the SPACAL calorimeter (scintillating fibers, lead absorber, sensitive to *all* ionizing particles, cf. Fig. 5.8). The profiles in the Čerenkov calorimeter are considerably narrower, even though both the $Z$ value and the density of the absorber in this calorimeter were considerably smaller than in the other one.

This figure illustrates one very important point, which is emphasized time and again in this book, namely that calorimeter signals are in general *not proportional* to the amount of energy deposited in the area from which they are collected. This becomes extremely clear in the case of the Čerenkov calorimeter. The profile measured with the Quartz Fiber detector is not at all representative for the energy deposit profile in the shower development. It just measures the transverse distribution of Čerenkov light produced in this process and thus the transverse distribution of shower particles capable of emitting such light. Since most of these particles are electrons and

**Fig. 5.15** Signal
distributions for 75 GeV
pions and electrons in a
preshower detector used in
beam tests of CDF
calorimeters [2]

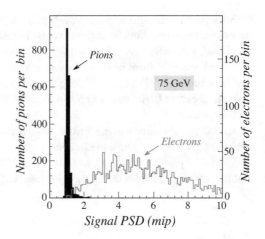

positrons generated in the em shower core, this light is concentrated near the shower
axis.

## 5.4  Application of Differences

The large difference between the radiation length and the nuclear interaction length in
high-$Z$ materials can be successfully exploited for electron/pion separation. In many
calorimeter systems, this difference manifests itself in the form of a very different
energy sharing between the longitudinal calorimeter sections, for showers induced
by electrons and pions, respectively.

However, the difference between (the very early phases of) em and hadronic
shower development may also offer excellent opportunities for $e/\pi$ separation in
structures that are very much thinner than a shower-absorbing calorimeter system.
This feature forms the basis of many *preshower detectors*. An extremely simple
preshower detector (PSD) may consist of a plate of lead, 1 cm (1.9 $X_0$, 0.06 $\lambda_{int}$)
thick, followed by a sheet of plastic scintillator. When a beam consisting of a mixture
of high-energy electrons and pions is sent through this device, almost all pions (96%)
traverse it without strongly interacting. These pions produce a minimum ionizing
peak in the scintillator. On the other hand, the electrons lose a considerable fraction
of their energy by radiating large numbers of bremsstrahlung photons. Some of these
photons convert into $e^+e^-$ pairs in the PSD and thus contribute to the scintillation
signals produced by this device.

The result is a very clear separation between electrons and pions. Figure 5.15
shows the signal distributions for 75 GeV electrons and pions in the described device,
used in beam tests of the CDF Plug Upgrade calorimeter [18, 19]. Even with such
simple devices, pion rejection factors of the order of hundred are readily achieved.

Such preshower detectors are particularly helpful during beam tests, for example for obtaining a pure sample of pion events. Especially at low energy, pion beams at accelerators are often contaminated with electrons. By placing a device of the type described above in front of the calorimeter to be tested, this electron contamination can be effectively eliminated, through a cut on the PSD signals. Since pions are defined as events that cause a mip signal in the PSD, the pion sample is not biased by this procedure.

Although the electron signals from the described PSD are substantial (compared with the signal from a mip), the fraction of the energy lost in this very early shower development phase is small. Provided that the PSD is placed directly in front of the calorimeter, the effect on the energy resolution of the latter is even smaller [20].

# References

1. Atwood, W.B., et al.: Astrophys. J. **697**, 1071 (2009)
2. Wigmans, R.: Calorimeters. In: Grupen, C., Buvat, I. (eds.) Handbook of Particle Detection and Imaging, vol. 1, p. 497. Springer, Berlin (2011)
3. Wigmans, R., Zeyrek, M.T.: Nucl. Instrum. Methods **A485**, 385 (2002)
4. Wigmans, R.: Calorimetry—Energy Measurement in Particle Physics, 2nd edn. International Series of Monographs on Physics, vol. 168. Oxford University Press, Oxford (2017)
5. Storm, E., Israel, H.I.: Nucl. Data Tables **7**, 565 (1970)
6. Bathow, G., et al.: Nucl. Phys. B **20**, 592 (1970)
7. Antonello, M., et al.: Nucl. Instrum. Methods **A899**, 52 (2018)
8. Akchurin, N., et al.: Nucl. Instrum. Methods **A533**, 305 (2004)
9. Abramowicz, H., et al.: Nucl. Instrum. Methods **180**, 429 (1981)
10. Acosta, D., et al.: Nucl. Instrum. Methods **A316**, 184 (1992)
11. Acosta, D., et al.: Nucl. Instrum. Methods **A309**, 143 (1991)
12. Barreiro, F., et al.: Nucl. Instrum. Methods **A292**, 259 (1990)
13. Leroy, C., Sirois, Y., Wigmans, R.: Nucl. Instrum. Methods **A252**, 4 (1986)
14. Dietrich, S.S., Berman, B.L.: At. Data Nucl. Data Tables **38**, 199 (1988)
15. Akchurin, N., et al.: Nucl. Instrum. Methods **A408**, 380 (1998)
16. Adragna, P., et al.: Nucl. Instrum. Methods **A615**, 158 (2010)
17. Akchurin, N., et al.: Nucl. Instrum. Methods **A399**, 202 (1997)
18. Apollinari, G., et al.: Nucl. Instrum. Methods **A409**, 547 (1998)
19. Albrow, M., et al.: Nucl. Instrum. Methods **A431**, 104 (1999)
20. Acosta, D., et al.: Nucl. Instrum. Methods **A308**, 481 (1991)

# Chapter 6
# The Energy Resolution of Calorimeters

## 6.1 Introduction

In the practice of a particle physics experiment, one will want to use a *given* calorimeter signal to determine the characteristics (e.g., the energy) of the particle that produced it. In order to be able to make a statement about the energy of a detected particle, one needs to know

1. the relationship between the measured signals and deposited energy (i.e., the detector calibration), and
2. the *energy resolution* of the calorimeter.

The energy resolution determines the *precision* with which the (unknown) energy of a given particle can be measured. It is experimentally determined from the precision with which the energy of particles of *known* energy is reproduced in the calorimetric measurements.

The energy resolution is often considered the most important performance characteristic of a calorimeter. In particle physics experiments, the energy resolution of the calorimeter may be the factor that limits the precision with which the mass of new particles can be determined (e.g., the top quark). It may limit the separation between particles with similar masses (e.g., in the jet–jet decay of the intermediate vector bosons $W$ and $Z$). And it determines the signal-to-background ratio in event samples collected in almost every experiment.

In this chapter, the factors that contribute to and limit the energy resolution of calorimeters are discussed. This resolution is determined by (*a*) the effects of fluctuations in the absorption process and (*b*) systematic effects. In Sect. 4.2, an example of a systematic effect was described, which was the result of the calibration procedure for the different sections of the calorimeter in question (Fig. 4.2). Such effects are typically ignored in physics analyses, but are nevertheless very real. At the end of this chapter, some other systematic effects that tend to be swept under the carpet are discussed. However, we start with the effects of fluctuations on the precision with which calorimetric measurements can be performed.

© Springer Nature Switzerland AG 2019
M. Livan and R. Wigmans, *Calorimetry for Collider Physics, an Introduction*,
UNITEXT for Physics, https://doi.org/10.1007/978-3-030-23653-3_6

## 6.2   The Effects of Fluctuations on the Calorimeter Performance

In the shower development that takes place when high-energy particles are absorbed in the block of matter that we call a calorimeter, the energy of the particles is degraded to the level of atomic ionizations or excitations that may be detected. The precision with which the energy of the showering particles can be measured is limited by

1. fluctuations in the processes through which the energy is degraded, and
2. the technique chosen to measure the final products of the cascade processes.

The fluctuations in the shower development process are unavoidable. In electromagnetic showers, they determine the ultimate limit on the achievable energy resolution. However, because of the chosen measurement techniques, the energy resolutions obtained in practice with em calorimeters are usually considerably worse than that.

The situation is quite different for hadron calorimeters. Contrary to em showers, in which the entire energy of the incoming particle is used to "heat up" the absorber, in hadron showers some fraction of the initial energy is used to break up atomic nuclei. This fraction, usually referred to as the *invisible* energy component since it does **not** contribute to the calorimeter signals, varies wildly from one event to the next. As a result, event-to-event variations in the calorimeter signal are much larger than for em showers of comparable energy, since this phenomenon has no equivalent in the latter. These fluctuations, and methods to mitigate their effects, are the topic of Chaps. 7 and 8.

Many, but not all, fluctuations contributing to the energy resolution of calorimeters obey the rules of Poisson statistics. For example, fluctuations in the number of quanta (scintillation or Čerenkov photons, ion–electron or electron–hole pairs, etc.) that constitute the calorimeter signals are Poissonian, but shower leakage fluctuations are not.

Let us assume that a particle with energy $E$ creates a signal $S$ that, on average, consists of $n$ signal quanta (e.g., photoelectrons). Event-to-event fluctuations in the signal correspond to Poisson (or Gaussian, for $n \gtrsim 20$) fluctuations in the number $n$. The relative width of the signal distribution, $\sigma_S / \langle S \rangle$, i.e., the relative precision of the calorimetric measurement of the energy, $\sigma_E / E$, is then equal to $\sqrt{n}/n = 1/\sqrt{n}$.

If the calorimeter is linear, it will produce a signal that consists, on average, of $4n$ quanta when it absorbs a particle with energy $4E$. The event-to-event fluctuations in the detection of such particles correspond to Gaussian fluctuations in the number $4n$. The relative precision of the calorimetric measurement of the energy of these particles amounts to $\sqrt{4n}/4n = 0.5/\sqrt{n}$, i.e., a factor of two ($= \sqrt{4}$) better than for particles with energy $E$.

For linear calorimeters measuring signal quanta that obey the rules of Poisson statistics, these considerations lead directly to the familiar relationship

$$\sigma_E / E = a / \sqrt{E} \tag{6.1}$$

We will follow the conventional practice of expressing the energy resolution of the calorimeter, $\sigma_E/E$, as a dimensionless number, representing a fraction of the particle energy $E$, e.g., 3.6%. Unless stated otherwise, we will assume that the particle energy is given in units of GeV. It has become customary to characterize calorimeters, for what concerns the precision with which they can measure the energy of the particles they absorb, in terms of the value of $a$. In this convention, $a$ thus represents the energy resolution for a 1 GeV energy deposit. Since calorimetry is based on statistical processes (the production of ionization charge, photons, electron–hole pairs, the excitation of Cooper pairs, etc.), the relative precision of the energy measurement ($\sigma_E/E$) thus *improves* with increasing energy. For example, an em calorimeter for which $a = 10\%$ will measure electrons of 4 GeV with a resolution of 5%, while the resolution will improve to 2% for 25 GeV electrons.

This very attractive feature has greatly contributed to the popularity of calorimeters in particle physics experiments. For other particle detection techniques, the relative precision of the measurements tends to deteriorate with increasing energy. This is most noticeably the case for momentum measurements in a magnetic field, where the size of the spectrometer has to increase proportional to $\sqrt{p}$ to keep the momentum resolution $\Delta p/p$ constant.

However, not all types of fluctuations contribute to the calorimetric energy resolution as $E^{-1/2}$. Some fluctuations are energy independent, e.g., fluctuations resulting from non-uniformities in the calorimeter structure. Other fluctuations may depend on the energy in a different way, e.g., fluctuations resulting from electronic noise ($E^{-1}$), or from lateral shower leakage ($E^{-1/4}$).

Not all types of fluctuations have a symmetric probability distribution around a mean value. As an example, we mention fluctuations in $f_{em}$, the em fraction of hadron showers. The probability of finding $f_{em}$ values larger than the most probable one is larger than that of finding smaller values (Fig. 7.2b).

In many cases, several sources of fluctuations contribute to the energy resolution of a given calorimeter. For example, Fig. 6.1 shows the energy resolution of the em barrel calorimeter of the ATLAS experiment as a function of energy, together with the various contributions to this energy resolution. Since their energy dependence may be different, the relative importance of each of these sources depends on the energy. For example, instrumental effects that cause energy-independent signal fluctuations tend to dominate at high energy, where the effects of Gaussian fluctuations has become very small. On the other hand, electronic noise, an important factor in LAr calorimeters, dominates the energy resolution at low energy, where its $E^{-1}$ dependence overtakes the $E^{-1/2}$ contributions from Gaussian fluctuations.

Most sources of fluctuations that contribute to a calorimeter's energy resolution are mutually uncorrelated. If that is the case, as in Fig. 6.1, the uncertainties they cause in the energy of the particles may be added in quadrature. This means that if sources 1, 2 and 3 cause fluctuations with standard deviations $\sigma_1$, $\sigma_2$ and $\sigma_3$ in the measurements of particles with energy $E$, then the total energy resolution amounts to $\sigma_E/E$, with

**Fig. 6.1** The em energy resolution and the separate contributions to it, for the em barrel calorimeter of the ATLAS experiment, at $\eta = 0.28$ [1]. Experimental data from [2]

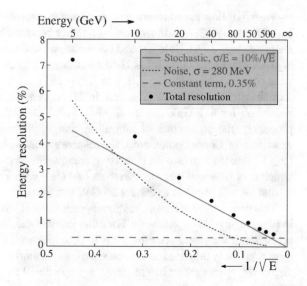

$$\sigma_E = \sqrt{(\sigma_1)^2 + (\sigma_2)^2 + (\sigma_3)^2} = \sigma_1 \oplus \sigma_2 \oplus \sigma_3 \qquad (6.2)$$

If the various sources are completely or partially correlated, their effects have to be combined accordingly. Depending on the details, this may result in resolutions that are either better or worse than expressed in Eq. 6.2.

The horizontal scale of Fig. 6.1 is one that is used as the standard throughout this book. The scale is linear in $E^{-1/2}$, with the origin located in the bottom right corner. In that way, the energy increases from left to right, as is common in plots that describe energy dependence, to reach $E = \infty$ for $E^{-1/2} = 0$. Since the horizontal axis is chosen in this way, experimental data that scale with $E^{-1/2}$ (Eq. 6.1) will be located on a straight line that extrapolates to the origin (0% resolution at infinite energy). The main reason for choosing this format is that it makes it immediately clear if and what type of deviations from $E^{-1/2}$ scaling are playing a role in the presented data. In general, the energy itself will be plotted on the top axis of plots of this type, which will further help to appreciate their contents.

## 6.3 Signal Quantum Fluctuations

Fluctuations in the number of detected signal quanta form the ultimate limit for the energy resolution that can be achieved with a given calorimeter. However, in most calorimeters, the resolution is dominated by other factors, discussed in the following sections. In this section, we describe examples of detectors in which fluctuations in the number of detected signal quanta determine the calorimeter resolution.

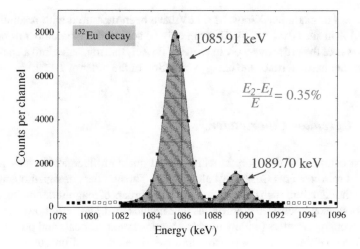

**Fig. 6.2** Detection of nuclear $\gamma$-rays, from the decay of $^{152}$Eu, with a high-purity germanium crystal. The energy resolution of this calorimeter is about 0.1% at 1 MeV. Courtesy of G. Roubaud, CERN

## 6.3.1 Semiconductor Crystals

The first example concerns the nuclear $\gamma$ detectors based on semiconductor crystals, such as Ge, Ge(Li) and Si(Li). It takes very little energy to create one electron–hole pair in these crystals, e.g., only 2.9 eV in germanium. The signal generated by a 1 MeV $\gamma$ fully absorbed in such a crystal therefore consists of some 350,000 electrons. The fluctuations in this number lead to an energy resolution of $1/\sqrt{350,000}$, or 0.17% (at 1 MeV!). In terms of Eq. 6.1, this means that $a = 0.005\%$, orders of magnitude smaller than anything we will see for "conventional" calorimeters used in particle physics experiments.

Owing to correlations in the production of consecutive electron–hole pairs (the so-called Fano factor [3, 4]), the limit on the energy resolution given by fluctuations in the number of primary processes is even smaller than indicated above. In practice, energy resolutions close to 1.0 keV at 1 MeV are indeed achieved with such detectors (Fig. 6.2).

## 6.3.2 Cryogenic Detectors

As a second example, some cryogenic detectors discussed in Sect. 1.4.4 can be mentioned. *Superconducting Tunnel Junctions* are based on the excitation of Cooper pairs, which then tunnel through a very thin layer separating two superconducting materials. This process takes less than 1 meV per Cooper pair. Therefore, small energy deposits may result in substantial numbers of primary processes and excellent energy resolution. For example, X-rays of 6 keV have been measured with resolutions of about 0.1% in such devices [5]. This can only be achieved if the number of primary processes is of the order of $\sim 10^6$. In terms of Eq. 6.1, this translates into a coefficient $a$ for the stochastic term of the energy resolution of the order of $10^{-6}$!

### 6.3.3 Čerenkov Calorimeters

The next example of detectors in which signal quantum fluctuations may play an important role concerns Čerenkov calorimeters. The numbers of signal quanta constituting the signals in such calorimeters are many orders of magnitude smaller than in the previous examples. Detectors of this type include lead-glass em shower counters, water Čerenkov counters (widely used in cosmic-ray experiments and proton-decay studies), as well as a variety of sampling calorimeters based on quartz as active medium. All these calorimeters are based on detection of Čerenkov light emitted by relativistic shower particles with velocities in excess of $c/n$.

In em showers, the signals are produced by relativistic electrons and positrons generated in the shower development. Typically, the Čerenkov threshold corresponds to a total energy of $\sim 0.7$ MeV for these particles (kinetic energy $>0.2$ MeV). Figure 3.9 indicates that in practice most of the em shower energy is deposited by particles capable of emitting Čerenkov light.

The Čerenkov mechanism accounts for an energy loss that is about 4 orders of magnitude smaller than the energy loss by ionization for superluminous charged particles (see Sect. 1.4.2). The spectrum of the emitted light has a characteristic $1/\lambda^2$ dependence, and thus has a dominating UV component. It is estimated that in media with $n \sim 1.5$ (e.g., water, quartz) the light yield in the visible part of the spectrum amounts to $\sim 30,000$ photons per GeV deposited energy [1]. The light yield in practical detectors is of course much smaller, because of light absorption, incomplete coverage by the light detectors, inefficiencies in the conversion of photons into photoelectrons (p.e.) and, in sampling calorimeters, the sampling fraction.

The largest light yield, $\sim 7,000$ p.e./GeV has been reported by Super-Kamiokande, which has a photocathode coverage of $\sim 40\%$ [6]. In lead-glass detectors, that coverage is much smaller and, therefore, only a small fraction of the Čerenkov photons produced are actually detected. Monte Carlo simulations lead to an estimate of $\sim 1,000$ p.e./GeV [1]. Statistical fluctuations in this number would lead to an expected energy resolution ($\sigma/E$) of $3.2\%/\sqrt{E}$.

The best lead-glass detector systems have reached energy resolutions of $\sim 5\%/\sqrt{E}$ for electromagnetic showers in the energy range from 1 to 20 GeV [7, 8]. This might indicate that Gaussian fluctuations in the number of photoelectrons are not the only factor determining the energy resolution of this type of calorimeter, although it is most definitely a very important factor. Additional factors that may play a role include, but are not limited to

- Fluctuations in the energy fraction deposited by shower particles capable of emitting Čerenkov light. On average, some 80% of the energy is deposited by particles with velocities above the Čerenkov threshold, but this fraction fluctuates from one event to the next. It depends, among other things, on the number of positrons produced in the shower development. This is because it takes more shower energy to produce a relativistic positron (through $\gamma \rightarrow e^+e^-$ conversion) than to accelerate an atomic electron to relativistic velocities (Compton scattering).
- Fluctuations in the spectrum of the particles capable of emitting Čerenkov light. Such fluctuations translate into fluctuations in the number of Čerenkov photons because of the strong $\beta$-dependence of the light yield ($\sin^2 \theta_C$, see Sect. 1.4.2).
- Fluctuations in the spectrum of the Čerenkov photons emitted by the relativistic shower particles. These fluctuations may be important since the quantum efficiency of the light detector is strongly wavelength dependent, with the cutoff occurring in the most densely populated wavelength area.

Another type of calorimeter based on the Čerenkov mechanism as the source of the signals uses quartz fibers as the sampling medium. In such calorimeters, the light yield is several orders of magnitude smaller than in the detectors mentioned above, because only a small fraction of the shower energy is deposited in the fibers. In addition, only light emitted within the numerical aperture of these fibers may contribute to the signals.

An example of such a detector is the quartz-fiber calorimeter installed in the very-forward region of the CMS experiment (HFCAL). The light yield of this detector is extremely small, less than 1 photoelectron (p.e.) per GeV. Because of the $1/\lambda^2$ dependence of the Čerenkov light intensity, the precise light yield (and thus the energy resolution) is affected by absorption in the windows of the PMTs that detect the light signals.

This is illustrated in Fig. 6.3, which shows results for electron showers measured with two different PMTs, one equipped with a glass window and the other with a quartz window, but otherwise identical. Since the latter window transmitted a larger fraction of the Čerenkov light, the signals were larger and the energy resolution was correspondingly better [9].

From the measured resolutions, one could actually infer the light yield of this detector. At 1 p.e./GeV, one expects the parameter $a$ in Eq. 6.1 to be 1.0 (100%), for 2 p.e./GeV, $a$ becomes $1/\sqrt{2}$ (71%) and for 0.5 p.e./GeV, $a = \sqrt{2}$ (141%). In general, the number of photoelectrons per GeV is given by $a^{-2}$ in this detector. In this way, the authors found a light yield of 0.87 p.e./GeV for the measurements with the PMT equipped with the quartz window and 0.53 p.e./GeV for the glass PMT. These numbers were in excellent agreement with the light yield found in dedicated measurements of this quantity.

One may conclude from these results that contributions to the energy resolution from factors other than fluctuations in the number of photoelectrons are negligible in this detector. The reason for this is the very small light yield, which causes large signal fluctuations, thus dwarfing the effects of other contributions (e.g., sampling fluctuations) to the energy resolution.

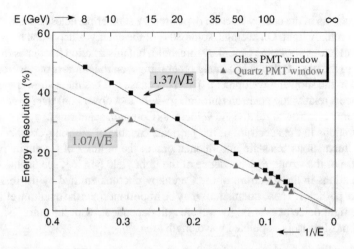

**Fig. 6.3** The energy resolution for electron detection with the QFCAL prototype detector, as a function of energy. Results are given for measurements in which photomultiplier tubes with a glass window were used and for measurements in which PMTs of the same type were equipped with a quartz window. From: Akchurin, N. et al. (1997). *Nucl. Instrum. Methods* **A399**, 202

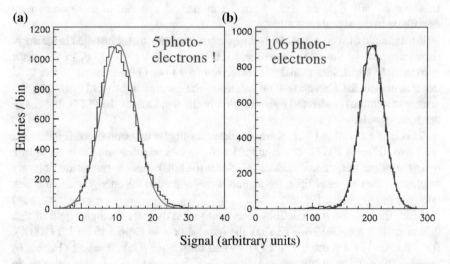

**Fig. 6.4** Signal distributions for 10 GeV (**a**) and 200 GeV (**b**) electrons showering in the CMS quartz-fiber calorimeter, measured with a PMT with a glass window. The curves represent Gaussian fits to the experimental data. From: Akchurin, N. et al. (1997). *Nucl. Instr. and Meth.* **A399**, 202

The light yield of this calorimeter is very small indeed. At 10 GeV, the measured signals were composed of 5.3 photoelectrons, on average, when read out with a PMT with a glass window. This small number of photoelectrons caused an asymmetric line shape (Fig. 6.4a), characteristic for a Poisson distribution $P_n$ for a discrete variable $n$ with a small average value $\mu$:

$$P_n = \frac{\mu^n}{n!} \, e^{-\mu} \tag{6.3}$$

At high energy, the number of photoelectrons was so large that the Poisson fluctuations led to a symmetric, Gaussian line shape. This is illustrated in Fig. 6.4b, for 200 GeV electrons, whose average signal consisted of 106 photoelectrons. The energy resolution $\sigma/E$ thus corresponded to $1/\sqrt{106}$, or 9.7%, for these high-energy electrons. It improved to $1/\sqrt{174}$ (7.6%) when the calorimeter was read out with a PMT with a quartz window.

### 6.3.4  Calorimeters Based on Scintillation

One type of detector for which the energy resolution is *not* limited by fluctuations in the number of signal quanta, is the scintillation counter. This may be illustrated by the following experimental data. Measurements with NaI(Tl) crystals on 6 keV X-rays have yielded a resolution $\sigma_E/E \approx 15\%$. If we assume that this result is dominated by fluctuations in the number of signal quanta, then this implies that the signals consist on average of $\sim 40$ photoelectrons. On the basis of this result, one would then for 1 MeV $\gamma$-rays expect resolutions of $15\%/(\sqrt{1{,}000/6}) \approx 1.2\%$. Yet, in reality the best resolutions obtained at this energy are only about 5%. In NaI(Tl) crystals, electromagnetic showers are detected with resolutions of about 1% at 1 GeV, whereas a factor of thirty ($\sqrt{1{,}000}$) improvement of the 1 MeV result should be expected if photoelectron statistics limited the resolution. Clearly, the energy resolution of this and other scintillation calorimeters is dominated by other factors.

When factors that are not determined by Poisson statistics contribute, the energy resolution does not scale as $E^{-1/2}$ with the energy of the showering particles. This is illustrated by Fig. 6.5, in which the data points do not extrapolate to the bottom right hand corner, as in Fig. 6.3. These data were measured with a sampling calorimeter, which was built in the context of the SPAKEBAB R&D project [10]. This detector had a sandwich structure. It consisted of a large number of very thin (0.63 mm thick) lead sheets, interleaved with 1 mm thick plastic scintillator plates. The scintillation light was transported to the rear end of the calorimeter by means of a large number of wavelength-shifting fibers, which were spaced by 4 mm and ran through the entire detector structure. Because of the very high sampling frequency, this calorimeter had an em energy resolution that was among the very best ever achieved with a sampling calorimeter.

It turned out that the contribution of Poisson fluctuations to the energy resolution ($5.7\%/\sqrt{E}$) was dominated by sampling fluctuations, and not by fluctuations in the number of scintillation photoelectrons. As a matter of fact, this is how it should be. Sampling fluctuations should always dominate in well designed sampling calorimeters based on scintillation or ionization charge collection. In Sect. 6.5, we describe how the contributions from sampling fluctuations and signal quanta fluctuations can be distinguished and measured.

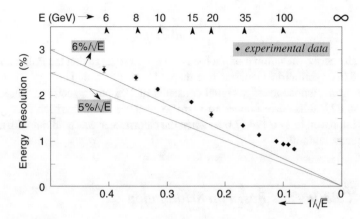

**Fig. 6.5** The energy resolution of the SPAKEBAB calorimeter (0.63 mm sampling layers) for electrons as a function of the electron energy. For comparison, the results for $\sigma/E = aE^{-1/2}$, with $a = 5\%$ and 6% (the straight lines) are given as well. From: Dubois, O. et al. (1996). *Nucl. Instr. and Meth.* **A368**, 640

## 6.4   Sampling Fluctuations

Sampling calorimeters, which are typically much cheaper than homogeneous ones, are especially competitive at higher energies. In properly designed instruments of this type, the energy resolution is determined by *sampling fluctuations*. These represent fluctuations in the number of different shower particles that contribute to the calorimeter signals, convolved with fluctuations in the amount of energy deposited by individual shower particles in the active calorimeter layers. They depend both on the *sampling fraction*, which is determined by the ratio of active and passive material, and on the *sampling frequency*, determined by the number of different sampling elements in the region where the showers develop. Sampling fluctuations are stochastic and their contribution to the energy resolution can be described by

$$(\sigma/E)_{\mathrm{samp}} = \frac{a_{\mathrm{samp}}}{\sqrt{E}}, \quad \text{with} \quad a_{\mathrm{samp}} = 0.027\sqrt{d/f_{\mathrm{samp}}} \tag{6.4}$$

in which $d$ represents the thickness of individual active sampling layers (in mm), and $f_{\mathrm{samp}}$ the sampling fraction for minimum ionizing particles (*mips*). This expression describes the em energy resolution obtained with a variety of different sampling calorimeters based on plastic scintillator or liquid argon as active material reasonably well (Fig. 6.6) [11].

Table 6.1 lists some characteristics of a representative selection of sampling calorimeters used in particle physics experiments. Above 100 GeV, the resolution of all calorimeters mentioned above is ∼1%, and systematic factors, such as stability of the electronic components, the effects of light attenuation, or temperature variations of the light yield, tend to dominate the performance.

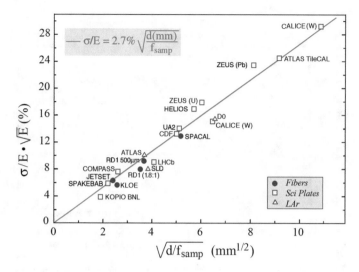

**Fig. 6.6** The em energy resolution of a variety of sampling calorimeters as a function of the parameter $(d/f_{samp})^{1/2}$, in which $d$ is the thickness (in mm) of an active sampling layer (e.g., the diameter of a fiber or the thickness of a liquid-argon gap), and $f_{samp}$ the sampling fraction for mips. The energy $E$ is expressed in units of GeV [11]

**Table 6.1** A representative selection of electromagnetic sampling calorimeters used in past and present particle physics experiments. The energy $E$ is expressed in GeV, the sampling fraction $f_{samp}$ refers to minimum ionizing particles. For the energy resolution, only the $E^{-1/2}$ scaling term, which dominates in the practically important energy range for these experiments, is listed

| Experiment | Calorimeter structure | $X_0$ (cm) | $f_{samp}$ (%) | $\sigma/E$ | References |
|---|---|---|---|---|---|
| KLOE (Frascati) | Pb/fibers | 1.6 | 17 | $4.7\%/\sqrt{E}$ | [12] |
| ZEUS (DESY) | $^{238}$U/scintillator | 0.7 | 9 | $18\%/\sqrt{E}$ | [13] |
| NA48 (CERN) | Pb/LKr | 1.5 | 23 | $3.5\%/\sqrt{E}$ | [14] |
| ATLAS (LHC) | Pb/LAr | $\approx 3$ | $\approx 25$ | $10\%/\sqrt{E}$ | [15] |
| PHENIX (RHIC) | Pb/scintillator | 3.1 | 29 | $7.8\%/\sqrt{E}$ | [16] |
| AMS-02 (ISS) | Pb/fibers | 1.3 | 19 | $10.4\%/\sqrt{E}$ | [17] |

The validity of Eq. 6.4 is limited to calorimeters with plastic scintillator or liquid argon/krypton as active material. When the active layers are very thin (in terms of stopping power), as in calorimeters with gaseous or silicon readout, an additional factor contributes to the energy resolution: *pathlength fluctuations*. In such calorimeters,

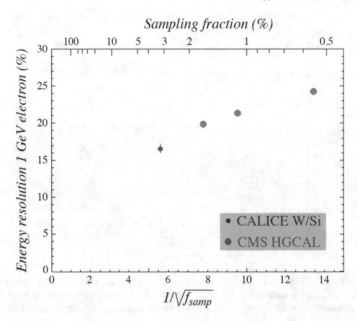

**Fig. 6.7** The em energy resolution of sampling calorimeters with silicon layers as active material. From: Wigmans, R. (2018). *J. Progr. Part. Nucl. Phys.* **103**, 109. Experimental data from [19, 20]

the energy deposited by a typical shower electron depends on its trajectory inside the active material. For example, the energy loss of electrons in 100 μm silicon amounts to ∼115 keV [18]. The signal from shower electrons with energies larger than 115 keV produced in Compton scattering or photoelectric effect therefore depends on the angle at which they traverse an active layer. The larger the angle with the shower axis, the larger the contribution of these particles to the signal. For 500 μm silicon, the same is true for shower electrons with energies larger than 330 keV. And since the sampling fluctuations are determined by fluctuations in the *total energy deposited* by the shower particles, and since these soft electrons are an important component of the developing em showers, these pathlength fluctuations are important for calorimeters with very thin active layers.

This is illustrated by Fig. 6.7, which shows the energy resolution for calorimeters with thin silicon layers as active material. All calorimeters have approximately the same structure. Absorber layers of tungsten with a thickness that increases with depth from 1.5 mm to 4.5 mm are interleaved with thin layers of silicon. In the four configurations of which the em energy resolution is displayed, the thickness of the silicon is 100 μm, 200 μm, 300 μm and 525 μm, respectively. These calorimeters thus have different sampling fractions, but the ratio $d/f_{samp}$ used in Eq. 6.4 is approximately the same. The experimental data point comes from CALICE [19], which used 525 μm silicon, and obtained a resolution of $16.5\%/\sqrt{E}$ (Eq. 6.4 gives ∼$12\%/\sqrt{E}$ for this configuration). The other data points concern GEANT4 simulation results for the HGCAL upgrade calorimeter for the CMS endcap region [20], which has about the same $d/f_{samp}$ value as the CALICE one, but uses thinner silicon layers. The expected energy resolution for this device ranges from $19.9\%/\sqrt{E}$ for 300 μm silicon to $24.3\%/\sqrt{E}$ for 100 μm silicon. These results, once again, illustrate the importance of the contribution of soft shower particles to the signals from sampling calorimeters, discussed earlier in Sect. 3.5.2.

## 6.5 Measuring the Contribution of Different Types of Fluctuations

Because of the very small light yield, fluctuations in the number of photoelectrons are almost always the dominating contribution to the em energy resolution of calorimeters that use Čerenkov light as the source of their signals. However, in sampling calorimeters based on scintillation or direct detection of the ionization charge, sampling fluctuations usually dominate, even though the contribution from fluctuations in the number of signal quanta cannot be neglected either. Both types of fluctuations contribute a term that scales as $E^{-1/2}$. In the following, we describe two examples of detectors for which the relative contributions of these two sources have been disentangled. The principle of the methods used in this procedure is in both cases to make a change in the experimental configuration that affects *only* the fluctuations one wants to investigate and measure the effects of that change.

### 6.5.1 SPAKEBAB

As shown in Fig. 6.5, the em energy resolution of this scintillation sampling calorimeter was measured to have a scaling term of $5.7\%/\sqrt{E}$ [10]. This term was caused by a combination of sampling fluctuations and fluctuations in the number of photoelectrons. The measurement of the contribution of fluctuations in the number of photoelectrons to the energy resolution of this detector was performed with neutral density filters that were installed in front of the PMTs that were used to collect the scintillation light. Such a filter reduced the number of photoelectrons by a known factor, $f$ ($f > 1$). This factor could be determined by simply measuring the ratio of the average calorimeter signals for a given beam of electrons with and without the filter. By reducing the amount of light, the energy resolution for a given type of showers (e.g., those caused by 10 GeV electrons), deteriorates. By measuring this *change in energy resolution* for showers of different energy and with different filters, the contribution of fluctuations in the number of photoelectrons to the energy resolution could be determined unambiguously.

If $x$ represents the average number of photoelectrons per GeV shower energy, the average number of photoelectrons from showers initiated by electrons of $E$ GeV equals $xE$. The standard deviation of fluctuations in this number is $\sqrt{xE}$, and therefore the contribution of photoelectron statistics to the relative precision of the energy measurement amounts to

$$(\sigma/E)_{\text{p.e.}} = 1/\sqrt{xE} \qquad (6.5)$$

Sampling fluctuations contribute to the measured energy resolution as

$$(\sigma/E)_{\text{samp}} = a_{\text{samp}}/\sqrt{E} \qquad (6.6)$$

By performing the measurements for low-energy electron showers, the contributions to the energy resolution from other sources, in particular from those contributing in an energy-independent way (e.g., instrumental effects), was made insignificant. In that case, the energy resolution measured *without* filters can be written as

$$(\sigma/E)_{\text{nofilter}} = \sqrt{a_{\text{samp}}^2/E + 1/xE} \qquad (6.7)$$

If now the number of photoelectrons is reduced by a factor $f$, the new energy resolution becomes

$$(\sigma/E)_{\text{filter}} = \sqrt{a_{\text{samp}}^2/E + f/xE} = \sqrt{(\sigma/E)_{\text{nofilter}}^2 + (f-1)/xE} \qquad (6.8)$$

When $f$ is known, $x$ can thus be derived from the degradation of the energy resolution. This method yields the most accurate results at low energy, where the contributions of non-stochastic processes to the energy resolution are small and the differences between the filtered and unfiltered energy resolutions large.

The measurements were performed with filters that reduced the number of photons by factors ($f$) of about three and ten, respectively, for electron showers of 6, 8, 10 and 15 GeV. By combining all results, it was found that this calorimeter produced $1{,}300 \pm 90$ photoelectrons per GeV deposited energy. Therefore, photoelectron statistics contributed $2.8\%/\sqrt{E}$ to the em energy resolution. Since the total em energy resolution of this detector was measured to be $5.7\%/\sqrt{E}$, sampling fluctuations contributed $5.0\%/\sqrt{E}$ ($5.0 \oplus 2.8 = 5.7$).

For sampling calorimeters, this is a high light yield, the result of a very efficient light collection. In many other scintillation calorimeters used in particle physics experiments, the light yield is typically one order of magnitude smaller. Therefore, the contribution of photoelectron statistics to the energy resolution of such calorimeters is typically $\sim 10\%/\sqrt{E}$.

### 6.5.2 ZEUS

The second example concerns the ZEUS Collaboration, who measured the contribution of sampling fluctuations to the electromagnetic and hadronic energy resolutions of their uranium/plastic-scintillator and lead/plastic-scintillator calorimeter proto-types [21].

These calorimeters consisted of 2.5 mm thick polystyrene-based scintillator sheets, alternating with either 3 mm thick $^{238}$U plates or 10 mm thick lead plates. In order to measure the effects of sampling fluctuations, the calorimeter was split into "two interleaved calorimeters." This is schematically illustrated in Fig. 6.8c. The signals from the odd-numbered scintillator layers were summed up to form "Signal $E_A$," while "Signal $E_B$" consisted of the summed signals from the even-numbered layers. In this way, the detector could be considered as two independent sampling calorimeters, A and B, embedded in one and the same instrument.

Technically, this was achieved by covering one side of each scintillator plate with black tape. For the odd-numbered plates, the light from the left-hand side of the scintillators was prevented from reaching the wavelength-shifting plate (and its associated PMT which produced "Signal $E_B$"). For the even-numbered plates, the light from the right-hand side of the scintillators was prevented from contributing to "Signal $E_A$."

For each shower, the signals from A and B were recorded and separate distributions were made of $E_A$, $E_B$, and the combinations $E_A + E_B$ ($E_{sum}$) and $E_A - E_B$ ($E_{diff}$). It is useful to define the following variables:

$$\sigma_A = \frac{\Delta E_A}{\langle E_A \rangle}, \quad \sigma_B = \frac{\Delta E_B}{\langle E_B \rangle}, \quad \sigma_{sum} = \frac{\Delta E_{sum}}{\langle E_{sum} \rangle}, \quad \sigma_{diff} = \frac{\Delta E_{diff}}{\langle E_{sum} \rangle}$$

which denote the normalized standard deviations of these four distributions.

The resolution of the complete calorimeter is then given by $\sigma_{sum}$, while $\sigma_{diff}$ measures the contribution of sampling fluctuations, $\sigma_{samp}$, to this total resolution. This is most easily understood if we consider em showers. In the approximation that only sampling fluctuations contribute to the em energy resolution of sampling calorimeters, one expects $\sigma_{sum} = \sigma_{samp}$ and $\sigma_A = \sigma_B = \sigma_{sum}\sqrt{2} = \sigma_{samp}\sqrt{2}$, since the sampling fraction in each of the two interleaved calorimeters A,B is a factor of two smaller than that of the total instrument. And if fluctuations in the signals from A and B are completely uncorrelated, $\Delta E_{sum} = \Delta E_{diff}$ and, therefore, $\sigma_{sum} = \sigma_{diff} = \sigma_{samp}$.

If other types of fluctuations contribute to the energy resolution, as in the detection of hadron showers, one will find $\sigma_{sum} > \sigma_{diff}$ and $\sigma_{A,B} < \sigma_{sum}\sqrt{2}$. In the extreme case, where the contribution of sampling fluctuations to the total resolution is negligible, $\sigma_A$ and $\sigma_B$ would be about equal to $\sigma_{sum}$, since sampling fluctuations are the only factor in which these two configurations differ. In general, the contributions of sampling fluctuations can be derived from the measured values of $\sigma_A$, $\sigma_{sum}$ and $\sigma_{diff}$ by assuming $\sigma_{samp}(A) = \sigma_{samp}(A+B) \times \sqrt{2}$.

In practice, the situation was slightly more complicated than described above, because fluctuations in the numbers of photoelectrons and effects of light attenuation in the scintillator plates also contributed measurably to the total energy resolution, especially for em showers. This meant that the difference in em resolution between calorimeter A and the total instrument was not only affected by the change in sampling fluctuations, but also by photoelectron statistics.

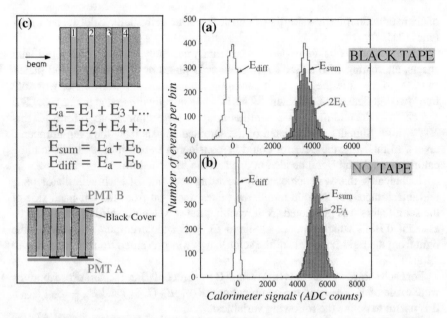

**Fig. 6.8** Pulse height distributions for 30 GeV hadrons obtained with the ZEUS lead/plastic-scintillator prototype calorimeter. Diagram **a** shows the distributions of $E_{sum}$, $E_{diff}$ and $2E_A$, measured in the configuration depicted in Fig. 6.8c, with the black tape in place. Diagram **b** shows the same distributions measured in the same configuration, but with the black tape removed. See text for details. From : Drews, G. et al. (1990). *Nucl. Instr. and Meth.* **A290**, 335

This contribution to the resolution was obtained by removing the black tape (see Fig. 6.8c) and repeating the measurements. In this new geometry, the two signals $E_A$ and $E_B$ resulting from a developing shower were thus measured with one and the same sampling device, but each signal contained only (approximately) half of the photoelectrons generated in the event. Therefore, $\sigma_{diff}$ measured in that case the fluctuations in the number of photoelectrons.

An example of the experimental results obtained in these measurements is shown in Fig. 6.8a, b. This figure contains pulse height distributions measured for 30 GeV hadrons with the 10 mm lead/2.5 mm plastic-scintillator prototype calorimeter. Each of the two diagrams shows three distributions: $E_{sum}$, $E_{diff}$ and $2E_A$. The latter distribution was obtained by multiplying the A signals by a factor of two and thus has the same central value as the distribution of the summed signals of A and B ($E_{sum}$). Diagram *a* shows the results obtained in the configuration depicted in Fig. 6.8c, with black tape, in diagram *b* the black tape was removed. This figure exhibits the following features:

- When the tape was removed, $\sigma_{diff}$ became considerably smaller. This means that fluctuations in the number of photoelectrons were much smaller than the sampling fluctuations.

- When the tape was removed, the distributions of $E_{\text{sum}}$ and $2E_A$ were practically identical. Since these two distributions only differ in the average number of photoelectrons constituting the signals, this means that fluctuations in the number of photoelectrons did not contribute significantly to the hadronic energy resolution of this calorimeter.
- When the black tape was in place, the distributions of $E_{\text{sum}}$ and $2E_A$ were *not* identical. In this configuration, the sampling fraction was different by a factor of two for these two distributions. Sampling fluctuations were, therefore, a major contribution to the hadronic energy resolution of this calorimeter.
- The distributions of $E_{\text{sum}}$ and $E_{\text{diff}}$ had practically the same width when the black tape was removed. This means that sampling fluctuations were not only a major contribution to the hadronic energy resolution, but that these fluctuations completely dominated the resolution. Less surprisingly, the same phenomenon was observed for the electron data.

The latter observations also reveals a very important other feature. If the widths of the distributions of $E_{\text{sum}}$ and $E_{\text{diff}}$ are essentially the same, then this implies that the signal distributions of $E_A$ and $E_B$ are completely *uncorrelated*. In both calorimeter configurations, the same showers were sampled, but the shower particles contributing to the signals in configuration A were not the same as those contributing to the signals in configuration B. Therefore, the typical shower particle contributing to the signals from this calorimeter traveled a distance that was short compared with the thickness of one sampling layer. The signal contribution from shower particles that traversed more than one sampling layer was negligible. This was earlier concluded by Willis and Radeka, who did similar measurements for electrons in their much-finer-sampling Fe/LAr calorimeter [22].

These measurements confirm that the signals from em calorimeters are dominated by electrons with energies that are much smaller than the critical energy, and that the signals from hadron showers are dominated by protons. Pions produced in the shower development, which could traverse many sampling layers, play indeed a very minor role.

After a careful analysis of all available experimental information, the ZEUS authors were able to unravel the em and hadronic energy resolutions of these two compensating calorimeters in their different contributing components. The results of this analysis are summarized in Table 6.2.

From these results, the following conclusions can be drawn:

1. The energy resolutions for em showers are strongly dominated by sampling fluctuations in these calorimeters, with a minor contribution coming from photoelectron statistics (increasing $a_{\text{samp}} = 16.5\%$ to $a_{\text{total}} = 18.5\%$ for the uranium detector, and $a_{\text{samp}} = 23.5\%$ to $a_{\text{total}} = 24.5\%$ for the lead detector). Given the size ot the error bars, all one can say is that the light yield is $\gtrsim 100$ p.e./GeV.
2. The sampling fluctuations for electrons are in good agreement with Eq. 6.4. For the uranium calorimeter, this formula predicted $a_{\text{samp}} = 15.6\%$, while the experimental value was measured to be $16.5 \pm 0.5\%$. For lead, the prediction of 21.9% was also close to the experimental value ($23.5 \pm 0.5\%$).

**Table 6.2** The contributions of sampling fluctuations and intrinsic fluctuations to the energy resolutions for electrons and pions in compensating uranium/plastic-scintillator and lead/plastic-scintillator calorimeters. Listed are the values of the coefficient $a$ (Eq. 6.1), expressed in %. Data from [21]

| Fluctuations (%) | 3 mm uranium/2.5 mm plastic | | 10 mm lead/2.5 mm plastic | |
| --- | --- | --- | --- | --- |
| | Electrons | Pions | Electrons | Pions |
| $\sigma_A, \sigma_B$ | $26.6 \pm 1.0$ | $49.5 \pm 1.0$ | $36.0 \pm 1.0$ | $60.5 \pm 1.0$ |
| $\sigma_{sum}$ | $18.5 \pm 1.0$ | $37.3 \pm 1.0$ | $24.5 \pm 1.0$ | $43.5 \pm 1.0$ |
| $\sigma_{diff}$ | $19.2 \pm 1.0$ | $32.6 \pm 1.0$ | $25.8 \pm 1.0$ | $42.3 \pm 1.0$ |
| $\sigma_{samp}$ | $16.5 \pm 0.5$ | $31.1 \pm 0.9$ | $23.5 \pm 0.5$ | $41.2 \pm 0.9$ |
| $\sigma_{intr}$ | $2.2 \pm 4.8$ | $20.4 \pm 2.4$ | $0.3 \pm 5.1$ | $13.4 \pm 4.7$ |

3. The energy resolutions for hadrons are also strongly dominated by sampling fluctuations in these calorimeters, especially in the lead one. As a result, the resolutions of calorimeters A and B are, *in all cases*, within experimental errors equal to the resolution of the complete instrument times $\sqrt{2}$.

4. The hadronic sampling fluctuations are about a factor of two larger than the em sampling fluctuations. This indicates that the number of different shower particles constituting the hadronic signal is much smaller. One reason for that is the fact that the specific ionization of the spallation protons that dominate the signals is much larger than for mips (Fig. 4.4a).

5. The contributions of "intrinsic" fluctuations, i.e., the fluctuations that remain after subtracting the contributions from sampling fluctuations, photoelectron statistics and other instrumental effects from the total measured hadronic energy resolution, scale with $E^{-1/2}$. This is typical for compensating calorimeters.

6. The intrinsic fluctuations are smaller in the lead calorimeter than in the uranium one.

Results 5 and 6 are further discussed in Chap. 8.

## 6.6   Shower Leakage

Calorimeters are instruments to measure the properties of particles by means of total absorption. However, in practice "total" means 99.9%, or 99%, or even less. When designing a calorimeter system for an experiment, and in particular for a $4\pi$ experiment, the choice of the calorimeter depth has important consequences, especially for the cost of the experiment. Therefore, it is important that the decision concerning the degree of shower containment be based on accurate information.

There are two aspects to this problem. First, incomplete shower containment leads to energy-dependent event-by-event fluctuations in the shower leakage, which affects the quality of the calorimetric information. Second, incomplete shower containment means that shower particles escape the calorimeter. These particles may cause signals in other detectors, e.g., the muon system, which may disturb the performance of these detectors.

### 6.6.1  Effects of Leakage on the Calorimetric Quality

Shower leakage is an energy-dependent effect. The fraction of the energy carried by the showering particle that is *not* deposited in the (fiducial) calorimeter volume depends not only on the particle's energy, but also on the type of particle. In a given calorimeter, electrons of a given energy are better contained than protons of the same energy, which are in turn better contained than pions, on average that is. Although one might intuitively think that the average shower leakage fraction *increases* with energy, this is not automatically true and several calorimeters offer examples of the opposite effect. The calorimetric quality of shower detectors is affected by three different types of shower leakage:

1. *Longitudinal* leakage. When people talk or worry about shower leakage, it is usually this type of leakage. Shower particles escape detection by emerging from the calorimeter's rear end. Considerations about longitudinal shower leakage often drive the design of the calorimeter system, since the depth of the instrumented volume strongly determines its cost. However, it should also be mentioned that some of this type of leakage is in practice unavoidable, since neutrinos as well as muons produced in the decay of pions and kaons will even escape very deep calorimeters.
2. *Lateral* leakage. Although the design of a calorimeter is often driven by considerations about longitudinal shower containment, in practice the effects of lateral shower leakage on the calorimeter performance usually dominate. To determine the particle's energy, the area surrounding the shower axis over which the calorimeter signals are integrated is typically limited to such an extent that lateral losses are easily an order of magnitude larger than the longitudinal ones. From this perspective, one might conclude that many calorimeters are probably deeper than necessary.
3. *Albedo*, a.k.a. backsplash, i.e., leakage through the front face. Of the three mentioned types of leakage, this is the only one that is fundamentally unavoidable, no matter how large the detector is made.

### 6.6.2  The Different Types of Leakage

At a given (average) leakage level, the effects of shower fluctuations on the energy resolution are much larger if the leakage occurs longitudinally than when the energy leaks out sideways.

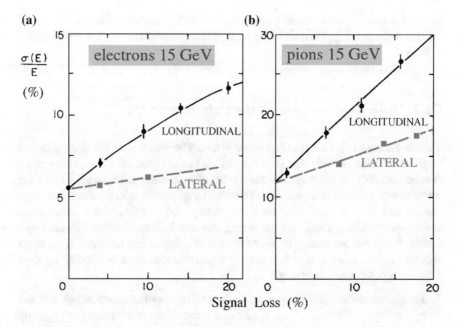

**Fig. 6.9** The effects of longitudinal and lateral shower leakage on the energy resolution, as measured for 15 GeV electrons (**a**) and pions (**b**) by the CHARM Collaboration in a low-$Z$ calorimeter [23, 24]. From: Amaldi, U. (1981). *Phys. Scripta* **23**, 409

This was already known a long time ago, as illustrated by Fig. 6.9. This figure shows the effects of longitudinal and lateral shower leakage fluctuations on the energy resolution of 15 GeV electrons (Fig. 6.9a) and pions (Fig. 6.9b), measured by the CHARM Collaboration for their low-$Z$ (marble absorber, $Z_{eff} \sim 13$) calorimeter [23, 24]. A 10% lateral leakage had a smaller effect on the energy resolution of this detector than a 5% longitudinal leakage, both for em and hadronic showers.

These differences can be qualitatively understood from the very different characteristics of longitudinal and lateral shower fluctuations. The longitudinal shower fluctuations, and thus the longitudinal leakage fluctuations, are for a very important part driven by fluctuations in the starting point of the shower, i.e., by the interaction characteristics of *one individual shower particle* (the first one). On the other hand, event-to-event fluctuations in the lateral shower development, and thus the fluctuations in lateral shower leakage, are determined by the interaction characteristics of $\sim 100$ *shower particles combined*. As a result, these fluctuations are much smaller than the longitudinal ones.

Shower leakage through the front face of the calorimeter (albedo) only plays a significant role at very low energies. The shower particles that escape in this way are, by definition, very soft, since they must be produced in scattering processes from which they emerged at large angles. Examples of such processes are Compton scattering and the photoelectric effect in the case of em showers and elastic neutron

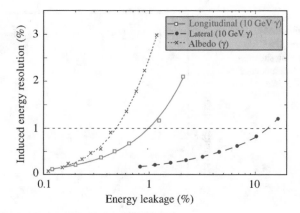

**Fig. 6.10** A comparison of the effects caused by different types of shower leakage. Shown are the induced energy resolutions resulting from albedo, longitudinal and lateral shower leakage as a function of the average energy fraction carried by particles escaping from the detector. The longitudinal and lateral leakage data concern 10 GeV $\gamma$s, the albedo data are for $\gamma$-induced showers of different (low) energies. Results from EGS4 Monte Carlo calculations in which the detector was represented by a block of tin. From: Wigmans, R. and Zeyrek, M.T. (2002). *Nucl. Instr. and Meth.* **A485**, 385

scattering in hadronic showers. Since the energies of the scattering products are at maximum a few MeV, only low-energy showers are significantly affected.

The effects of the different types of shower leakage on the em energy resolution are compared in Fig. 6.10. The effects of lateral shower leakage on the energy resolution are much smaller than for longitudinal leakage. According to Fig. 6.10, as much as 10% leakage can be tolerated before the induced energy resolution resulting from lateral leakage *fluctuations* exceeds 1%. A similar limitation would require 99% longitudinal containment.

The relative insensitivity of the em calorimeter resolution to lateral leakage is quite fortunate, since em showers tend to have considerable radial tails. This is illustrated by Fig. 5.1b, which shows the average lateral leakage fraction as a function of the radius of an infinitely deep calorimeter centered on the shower axis. It turns out that a radius of two Molière radii ($\rho_M$) is more than adequate for 90% lateral containment. This number is almost independent of the shower energy and the absorber material, in stark contrast with the depth requirements. However, in order to capture 99% instead of 90% of the shower energy, the detector mass has to be increased by an order of magnitude.

The SPACAL Collaboration has made an extensive study of various aspects of the shower leakage issue, for hadron-induced showers. Their lead/scintillating-fiber calorimeter measured $9.5\lambda_{int}$ in depth, with a diameter of $\sim 4.7\lambda_{int}$.

The longitudinal leakage from the back of this calorimeter was on average less than 0.5%, even for the highest-energy pions with which the calorimeter was tested (150 GeV). When the signals from all (155) readout cells were summed together,

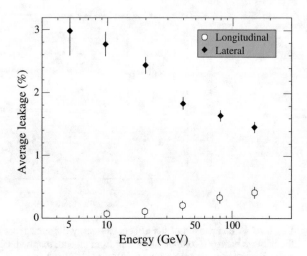

**Fig. 6.11** The average lateral and longitudinal leakage out of the SPACAL calorimeter, for pion showers as a function of beam energy [1]. Experimental data from [25, 26]

the lateral leakage was found to be four times as large as the longitudinal one, at 150 GeV (Fig. 6.11). At lower energies, the ratio between lateral and longitudinal shower leakage rapidly increased, to reach a factor of thirty at 10 GeV [25, 26].

It is interesting to note that the average fraction of the shower energy that leaked out laterally *increased* at lower energies. This means that high-energy pion showers were thus better contained than low-energy ones in this calorimeter (see also Sect. 5.3.3). This is a direct consequence of the increase of the (average) energy fraction contained in the em shower component, $\langle f_{em} \rangle$, with energy. At 10 GeV, this core represents on average ∼30% of the shower energy, rising to ∼60% at 150 GeV.

Shower leakage had a relatively larger impact on the calorimeter resolution as the energy increased. It turned out that the effect of lateral shower leakage on the hadronic energy resolution was reasonably described by a $E^{-1/4}$ term, added in quadrature to the stochastic term ($a_1 E^{-1/2}$) that described the calorimeter resolution in the absence of shower leakage:

$$\sigma/E = \sqrt{\left(\frac{a_1}{\sqrt{E}}\right)^2 + \left(\frac{x}{\sqrt[4]{E}}\right)^2 + \dots} \tag{6.9}$$

where $x$ represents the average lateral leakage fraction. More detailed results of these leakage studies can be found in [25, 27]. The essence of the conclusions about the effects of incomplete shower containment described here was also confirmed by studies that were performed with the ATLAS TileCal [28, 29].

## 6.7 Instrumental Effects

In real life, neither the calorimeter construction nor the environment in which it has to operate is ideal, and this has consequences for the calorimeter performance. In this section, we investigate the consequences of some common effects on the energy resolution.

The effects that are discussed have in common that the associated fluctuations do not scale with $E^{-1/2}$. This means that their relative contribution to the total energy resolution is energy dependent. Most effects, namely those that cause energy-independent fluctuations, are dominating the energy resolution at very high energies, where the contributions from the processes governed by Poisson statistics are small. However, the first effect to be discussed is an exception to this rule, since it dominates the resolution at low energies.

### 6.7.1 Electronic Noise

For sampling calorimeters with an active medium based on direct collection of the charge produced in the ionization-chamber (e.g., liquid argon) or proportional mode (wire chambers), the signals typically amount to a few picocoulombs of charge per GeV of shower energy.

The signals produced by these calorimeters correspond to the charge collected during a certain time, which we will call the "gate time" in the following. Since the detector has a certain capacitance, there is inevitably a contribution of electronic noise to the signals. This means that, in the absence of a showering particle, the integrated charge collected during the gate time fluctuates from event to event. Since the standard deviation of the electronic noise fluctuations corresponds to a certain, fixed energy, the contribution of this noise to the energy resolution of calorimetric shower measurements, $\sigma/E$, scales like $E^{-1}$.

The stochastic fluctuations are completely uncorrelated to the noise and the noise in the various electronic channels is completely incoherent (uncorrelated). Therefore the total energy resolution can be written as a quadratic sum of the two terms

$$(\sigma/E)_{\text{tot}} = \frac{a_{\text{noise}}}{E} \oplus \frac{a_{\text{stoch}}}{\sqrt{E}} \qquad (6.10)$$

Because of its $E^{-1}$ dependence, the noise term dominates the energy resolution at low energy. This is illustrated in Fig. 6.1.

Although these effects mainly play a role in calorimeters based on direct collection of the ionization charge, they may affect other types of calorimeters, in particular scintillator-based ones, as well. In scintillator calorimeters, the PMT signals are digitized and analyzed by means of analog-to-digital converters (ADCs). When no PMT signals are offered, the ADC may still accumulate a certain amount of charge during the gate time, resulting in a "pedestal." To find the calorimeter signals, the pedestal has to be subtracted from the raw signals.

However, the pedestal may exhibit fluctuations, e.g., due to ground loops, improper impedance matching and a variety of other electronic problems. These pedestal fluctuations play the same role for scintillator calorimeters as electronic noise does for ionization calorimeters. The difference is that pedestal fluctuations can in general be made insignificant, for example by increasing the gain of the PMTs, and thus the size of the signals. In liquid-argon calorimeters, which are based on the collection of *non-amplified* ionization charge, this option does not exist.

### 6.7.2  Variations in Sampling Fraction

The second source of instrumental effects that may contribute to the energy resolution may manifest itself in a variety of different ways, depending on the type of calorimeter. They all have the same common origin: event-to-event fluctuations in the sampling fraction (and thus the response) of the calorimeter volume in which the shower develops. These fluctuations typically lead to an energy independent contribution to the energy resolution. Examples of such effects include:

- Dependence of the sampling fraction on the impact point of the particle. This phenomenon plays a role, for example, in the "accordion" Pb/LAr ECAL of the ATLAS experiment, where it is an inherent consequence of the way in which active and passive elements have been arranged [30]). In Fig. 6.12, the average

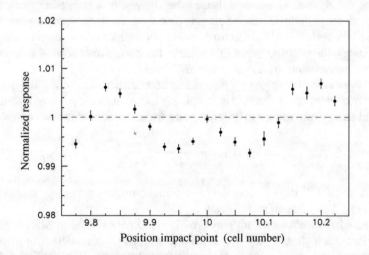

**Fig. 6.12**  The em response as a function of the impact position of the particles, for 90 GeV electrons in a prototype Pb/LAr accordion calorimeter for the ATLAS experiment. From: ATLAS (1996). *The ATLAS Calorimeter Performance*, report CERN/LHCC/96-40

calorimeter signal from 90 GeV electrons is plotted as a function of the impact point. The observed pattern is responsible for a constant term in the em energy resolution of ~0.4%. Similar effects have been reported for fiber calorimeters, when the particles enter at very small angles with the fiber axis [31]. SPACAL has shown that the constant term in the energy resolution can be avoided if the impact point of the particles can be measured with sufficient precision [27]. However, since the response fluctuations take place on a scale of (a fraction of) 1 mm, this requires an extremely good position resolution.

- Another type of position dependence arises from engineering considerations. In practice, calorimeters are built according to some modular structure, which has to be supported. The signals have to be transported to the outside world to be measured and analyzed, and utilities (power, cooling, etc.) have to be delivered to the detector from the outside world. As a result, there are usually "cracks" in a calorimeter system, regions that are not (or only partially) sensitive to particles developing showers. Mitigating, or minimizing the problems associated with this is usually a major aspect of designing a calorimeter system for a particular experiment. A lot of information about the optimal ways to incorporate "dead" material in a calorimeter structure is given in [1].

- One of the most serious mistakes that can be made in the design of a calorimeter is non-uniformity in depth. It may be financially attractive to let the sampling fraction depend on the (average) energy fraction deposited in the area in question, and thus gradually change the sampling fraction with depth. However, if one wants to subdivide the calorimeter into longitudinal segments with different sampling fractions, it is *absolutely crucial* to record the signals from these segments separately. Failure to do so will lead to a situation where fluctuations in the longitudinal shower development, more than anything else, determine the energy resolution [1]. The underlying reason for this is the same as the reason for the large effects of longitudinal shower leakage on the energy resolution (Sect. 6.6.2): the fluctuations are caused by only one shower particle.

### 6.7.3 Non-uniformity of Active Elements

Non-uniformities in the active elements also lead to a position dependent sampling fraction and thus cause fluctuations in the calorimeter response which are, in first approximation, energy independent and thus contribute a constant term to the energy resolution. These non-uniformities are often *avoidable*. They are closely linked to the tolerances maintained in the construction of the detector. Their effects can be minimized by proper quality control. Examples of such effects include:

- In liquid-argon calorimeters, non-uniformities may result from variations in the gap width. Such variations may occur over the surface of individual gaps, or from one gap to another. Since liquid-argon calorimeters operate in an ionization-chamber mode, the charge collected from an ionizing particle crossing a gap is directly proportional to the gap width. If a gap with a nominal width of 5 mm varies between 4.5 and 5.5 mm, then the signals from shower particles crossing this gap fluctuate by ±10%.

- In scintillation calorimeters, variations in the thickness of the scintillator plates or the scintillating fibers have the same effect for these calorimeters as the gap width in the previous case. In addition, the quantum efficiency for the conversion of light into photoelectrons may depend on the position where the photons hit the PMT's photocathode. And then there are the effects of light attenuation.
- Light attenuation may be caused by a variety of factors, of which self-absorption and reflection losses are the most common ones. Light attenuation causes the signals to be dependent on the distance the light has traveled between the position where it was generated and the position where it is converted into an electric signal, and thus causes a position dependence in the calorimeter response.

The light attenuation characteristics of all known scintillators change with time, for example as a result of chemical processes induced by the scintillation light itself [32], but in particular when the scintillator is operated in a radiation field. The mentioned position dependence of the signals is thus, in addition, a function of time.

There are various ways to deal with the effects of light attenuation. Many experiments, including ZEUS [33], HELIOS [34] and KLOE [35] have exploited the advantages of two-sided scintillator readout in their calorimeters and thus strongly mitigated the effects of light attenuation. Especially ZEUS went to great lengths to make the response of their uranium/plastic-scintillator calorimeter as uniform as possible.

In scintillating-fiber calorimeters, in which the fibers are oriented at 0°, with the PMTs located at the downstream end of the calorimeter, a similar increase in the effective attenuation length can be achieved by mirroring the upstream fiber ends. For individual scintillating fibers, an increase of the effective attenuation length from 6.8 m to 15.9 m was reported as a result of this procedure [36]. The SPACAL Collaboration combined this mirror technique with the application of filters that selectively absorbed the most attenuated components of the scintillation light, and increased the effective attenuation length of the fibers in their calorimeter to ~11 m in this way [27]. Despite this long attenuation length, the effects on the calorimeter signals were still evident. This is illustrated in Fig. 6.13, which shows a scatter plot of the signals from 150 GeV pions showering in SPACAL versus the average depth at which the light production took place [37]. Once the depth of the light production is known, one can trivially correct the measured signals for the effects of light attenuation. There are several ways to determine this depth event by event in longitudinally segmented calorimeters such as this one. One of these methods, which makes use of the time structure of the signals, is discussed in Sect. 8.5.

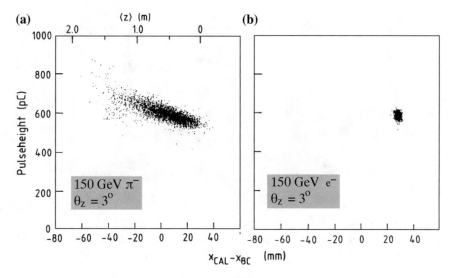

**Fig. 6.13** Scatter plot showing the SPACAL signal for 150 GeV $\pi^-$ (**a**) and 150 GeV $e^-$ (**b**) versus the center of gravity of the light production in the showers. The bottom scale shows the lateral displacement of the shower's center of gravity with respect to the particle's impact point. In the top scale, this displacement is converted into the average depth $\langle z \rangle$ of the light production. From: Acosta, D. et al. (1991). *Nucl. Instr. and Meth.* **A305**, 55

## 6.8  Misconceptions Affecting the Measured Energy Resolution

As stated in the beginning of this chapter, the energy resolution of a calorimeter is typically determined by measuring the signal distributions for beams of monoenergetic particles. By using beams with energies that span a large energy range, one can at the same time also establish the linearity of the calorimeter. We want to finish this chapter by mentioning a few misconceptions about the meaning of the concept of energy resolution. These lead to incorrect results and make it hard, if not impossible, to compare the (reported) performance of different calorimeters in an objective way.

- *The line shape.* Often, the measured signal distributions exhibit non-Gaussian tails. In that case, one should quote the $\sigma_{rms}$ value as the energy resolution. However, some authors use another variable, in order to make the results less dependent on the tails of the signal distributions they measure, and thus look better. This variable, called rms$_{90}$, is defined as the root-mean-square of the values located in the smallest range of reconstructed energies that contains 90% of the total event sample (see, e.g., Fig. 6.14). For the record, it should be pointed out that for a Gaussian distribution, this variable gives a 21% smaller value than $\sigma_{rms}$ (i.e., $\sigma_{fit}$). Of course, one is free to define variables as one likes. However, one should then not use the term "energy resolution" for the results obtained in this way, and compare results obtained in terms of rms$_{90}$ with genuine energy resolutions

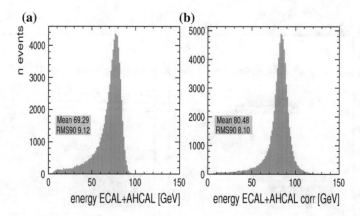

**Fig. 6.14** The line shape of the CALICE W/Si + Fe/plastic combination for 80 GeV pion showers, before (**a**) and after (**b**) correction procedures were applied based on the starting point of the showers and the estimated leakage. From: Sefkow, F. et al. (2016). *Rev. Mod. Phys.* **88**, 015003

from calorimeters with Gaussian response functions. This misleading practice is followed by the proponents of Particle Flow Analysis [38].

- *Saturation effects* in the sensors reduce the width of the signal distribution. Since a source of fluctuations is suppressed/eliminated because of this, the energy resolution of such calorimeters is in reality worse than suggested by the width of the signal distribution. This phenomenon affects, for example, the measurements of calorimeters in which the light signals are sensed by SiPMs. An extreme case of saturation occurs in digital calorimeters such as the one discussed in Sect. 4.2.2. Saturation effects go hand in hand with non-linearity. One may use weighting schemes to restore a semblance of linearity [39], but such schemes are of no consequence for the mis-measured energy resolution, since they do not address the fluctuations about the mean signal values that are modified in such procedures.

- *Non-linearity.* In general, one would want to know the energy resolution of a calorimeter over a certain range of energies. Even if the width of the signal distribution is determined in a statistically correct way, and if saturation effects do not play a role, the assumption that this width represents the energy resolution is only correct if the average value of that measured signal distribution corresponds indeed to the correct energy of the particles. As discussed in Sect. 4.2, signal non-linearities tend to invalidate that assumption.

- *Biased event samples.* One of the most common mistakes that are made when analyzing the performance of a calorimeter derives from the selections that are made to define the experimental data sample. This selection process may easily lead to biases, which distort the performance characteristics one would like to measure. In some extreme cases, this may lead to very wrong conclusions, such as the claim that uranium/liquid-argon calorimeters are compensating [1]. As a general rule, the calorimeter data themselves should *not* be used to apply cuts and thus select the event sample to be analyzed. Any such cuts should be based on

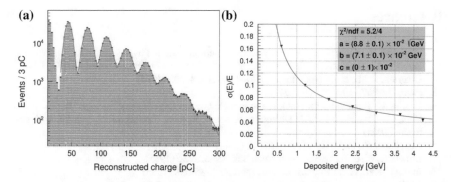

**Fig. 6.15** Distribution of the total collected charge in beam tests of a prototype calorimeter for the NA62 experiment at CERN (**a**). These tests were carried out with a beam of 606 MeV electrons. The peaks are the result of several beam particles entering the detector simultaneously. The energy resolution derived from these measurements (**b**). The data points were fit with a curve of the type $\sigma/E = aE^{-1/2} \oplus bE^{-1} \oplus c$. The resulting values of the coefficients $a$, $b$ and $c$ are shown in the legend. From: Antonelli, A. (2018). *Nucl. Instr. and Meth.* **A877**, 178

external detectors, such as upstream Čerenkov counters and/or preshower detectors, downstream leakage detectors and muon counters located behind substantial amounts of absorber material. Almost every analysis of test beam data we know of suffers from bias problems, the question is only *to what extent* the obtained results are affected by this.

- *Misinterpretation of experimental data.* Another mistake that is not uncommon concerns the extrapolation of measurement results far beyond their region of validity. A classic example concerned the "determination" of the energy resolution for high-energy em shower detection in liquid xenon, based on convolving the signals from large numbers of low-energy electrons (100 keV) recorded in a small cell [40]. These measurements only revealed something about the energy resolution for the detection of these low-energy electrons. In a high-energy em shower, a variety of new effects, absent or negligible in the case of these electrons, affect the signals and their fluctuations.

  A more recent example concerns beam tests of a calorimeter intended to detect $\gamma$s with energies up to 5 GeV at small scattering angles in the NA62 experiment [41]. The authors tested this detector in a 606 MeV electron beam. They obtained energy resolutions at higher energies by using signals in which $n$ such electrons entered the calorimeter simultaneously (Fig. 6.15). As before, such tests only reveal something about 606 MeV electrons.

  It should be stressed that, as a matter of principle, measurements made for low-energy particles cannot be used to determine/predict the high-energy calorimeter performance.

# References

1. Wigmans, R.: Calorimetry—Energy Measurement in Particle Physics, 2nd edn. International Series of Monographs on Physics, vol. 168. Oxford University Press, Oxford (2017)
2. Gingrich, D.M., et al.: Nucl. Instrum. Methods **A364**, 290 (1995)
3. Fano, U.: Phys. Rev. **72**, 26 (1947)
4. Leo, W.R.: Techniques for Nuclear and Particle Physics Experiments. Springer, Berlin (1987)
5. Hess, C.: Cryogenic Particle Detection. Topics in Applied Physics Series, vol. 99. Springer, Berlin (2005)
6. Abe, K., et al.: Phys. Rev. D **83**, 052010 (2011)
7. Brown, R.M., et al.: IEEE Trans. Nucl. Sci. **NS–32**, 736 (1985)
8. Jeffreys, P.W., et al.: Rutherford Lab report RAL-85-058 (1985)
9. Akchurin, N., et al.: Nucl. Instrum. Methods **A399**, 202 (1997)
10. Dubois, O., et al.: Nucl. Instrum. Methods **A368**, 640 (1996)
11. Livan, M., Vercesi, V., Wigmans, R.: Scintillating-fibre Calorimetry, CERN Yellow Report, CERN 95–02, Genève, Switzerland (1995)
12. Antonelli, A., et al.: Nucl. Instrum. Methods **A354**, 352 (1995)
13. Behrens, U., et al.: Nucl. Instrum. Methods **A289**, 115 (1990)
14. Barr, G.D., et al.: Nucl. Instrum. Methods **A370**, 413 (1996)
15. Aharouche, M., et al.: Nucl. Instrum. Methods **A568**, 601 (2006)
16. Kistenev, E.P.: In: Gordon, H., Rueger, D. (eds.), Proceedings of 5th International Conference on Calorimetry in High Energy Physics, Brookhaven Nat. Lab., p. 211. World Scientific, Singapore (1995)
17. Gallucci, G.: J. Phys. Conf. Ser. **587**, 012028 (2015)
18. Berger, M.J.: Stopping Power and Range Tables for Electrons, Protons and Helium Ions, report NISTIR 4999 (1993). http://physics.nist.gov/Star
19. Sefkow, F., et al.: Rev. Mod. Phys. **88**, 015003 (2016)
20. Collaboration, C.M.S.: Technical proposal for the phase-II upgrade of the Compact Muon Solenoid, CERN-LHCC-2015-10, CERN, Geneva (2015)
21. Drews, G., et al.: Nucl. Instrum. Methods **A290**, 335 (1990)
22. Willis, W.J., Radeka, V.: Nucl. Instrum. Methods **120**, 221 (1974)
23. Diddens, A.N., et al.: Nucl. Instrum. Methods **178**, 27 (1980)
24. Amaldi, U.: Phys. Scripta **23**, 409 (1981)
25. Acosta, D., et al.: Nucl. Instrum. Methods **A309**, 143 (1991)
26. Acosta, D., et al.: Nucl. Instrum. Methods **A316**, 184 (1992)
27. Acosta, D., et al.: Nucl. Instrum. Methods **A308**, 481 (1991)
28. Simonyan, M.: Performance of the ATLAS Tile Calorimeter to pions and protons. CERN-THESIS-2008-032 (2008)
29. Adragna, P., et al.: Nucl. Instrum. Methods **A615**, 158 (2010)
30. ATLAS: The ATLAS Calorimeter Performance, report CERN/LHCC/96-40 (1996)
31. Cardini, A., et al.: Nucl. Instrum. Methods **A808**, 41 (2016)
32. Sirois, Y., Wigmans, R.: Nucl. Instrum. Methods **A240**, 262 (1985)
33. Andresen, A., et al.: Nucl. Instrum. Methods **A309**, 101 (1991)
34. Åkesson, T., et al.: Nucl. Instrum. Methods **A262**, 243 (1987)
35. Babusci, D., et al.: Nucl. Instrum. Methods **A332**, 444 (1993)
36. Hartjes, F.G., Wigmans, R.: Nucl. Instrum. Methods **A277**, 379 (1989)
37. Acosta, D., et al.: Nucl. Instrum. Methods **A305**, 55 (1991)
38. Thomson, M.A.: Nucl. Instrum. Methods **A611**, 25 (2009)
39. Wigmans, R.: J. Progr. Part. Nucl. Phys. **103**, 109 (2018)
40. Séguinot, J., et al.: Nucl. Instrum. Methods **A323**, 583 (1992)
41. Antonelli, A.: Nucl. Instrum. Methods **A877**, 178 (2018)

# Part III
# Hadron Calorimetry: The Problems and How to Solve Them

# Chapter 7
# The Fundamental Problems of Hadron Calorimetry

## 7.1 Introduction

The development of hadronic cascades in dense matter differs in essential ways from that of electromagnetic ones, with important consequences for calorimetry. Hadronic showers consist of two distinctly different components:

1. An *electromagnetic* component; $\pi^0$s and $\eta$s generated in the absorption process decay into $\gamma$s which develop em showers.
2. A *non-electromagnetic* component, which combines essentially everything else that takes place in the absorption process.

For the purpose of calorimetry, the main difference between these components is that some fraction of the energy contained in the non-em component does *not* contribute to the signals. This *invisible energy*, which mainly consists of the binding energy of nucleons released in the numerous nuclear reactions, may represent up to 40% of the total non-em energy, with large event-to-event fluctuations. For this reason, homogeneous hadron calorimeters offer no particular advantage, as illustrated by Fig. 1.4. As a matter of fact, sampling calorimeters hold all the records for best hadronic performance.

In this chapter, we investigate the consequences of the fluctuations in invisible energy for the performance of hadron calorimeters.

## 7.2 The $e/h$ Ratio and Its Consequences

Let us define the calorimeter *response* as the conversion efficiency from deposited energy to generated signal, and normalize it to minimum ionizing particles (*mip*s). If the average signal is proportional to the deposited energy, i.e., if the calorimeter is linear, this definition implies that the response of this calorimeter is energy independent. The responses of a given calorimeter to the em and non-em hadronic

© Springer Nature Switzerland AG 2019    143
M. Livan and R. Wigmans, *Calorimetry for Collider Physics, an Introduction*,
UNITEXT for Physics, https://doi.org/10.1007/978-3-030-23653-3_7

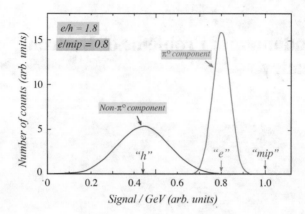

**Fig. 7.1** Illustration of the meaning of the $e/h$ and $e/mip$ values of a calorimeter. Shown are distributions of the signal per unit deposited energy for the electromagnetic and non-em components of hadron showers. These distributions are normalized to the response for minimum ionizing particles ("$mip$"). The average values of the em and non-em distributions are the em response ("$e$") and non-em response ("$h$"), respectively [1]

shower components, which we will call $e$ and $h$, are usually not the same, as a result of invisible energy and a variety of other effects. We will call the distribution of the signal per unit deposited energy around the mean value (i.e., $e$ or $h$ or the response to showers) the *response function*.

Figure 7.1 illustrates the different aspects of the calorimeter response schematically. The em response is larger than the non-em one, and the non-em response function is broader than the em one, because of event-to-event fluctuations in the invisible energy fraction. Both $e$ and $h$ are smaller than the calorimeter response for minimum ionizing particles, because of inefficiencies in the shower sampling process [2]. The calorimeter is characterized by the $e/h$ and $e/mip$ ratios, which in this example have values of 1.8 and 0.8, respectively. Calorimeters for which $e/h \neq 1$ are called *non-compensating*.

The properties of the em shower component have important consequences for the hadronic *energy resolution*, signal *linearity* and *response function*. The average fraction of the total shower energy contained in the em component, $\langle f_{em} \rangle$, was measured to increase with energy following a power law [3, 4], confirming an induction argument made to that effect [5]:

$$\langle f_{em} \rangle = 1 - \left[ \left( \frac{E}{E_0} \right)^{k-1} \right] \qquad (7.1)$$

where $E_0$ is a material-dependent constant related to the average multiplicity in hadronic interactions (varying from 0.7 GeV to 1.3 GeV for $\pi$-induced reactions on Cu and Pb, respectively), and $k \sim 0.82$ (Fig. 7.2a).

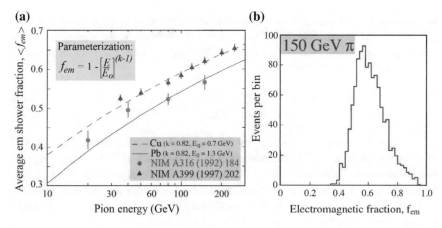

**Fig. 7.2** Properties of the electromagnetic fraction of hadron showers. Shown are the results of measurements of the average value of that fraction as a function of energy, for showers developing in lead or copper (**a**) [1] and the distribution of $f_{em}$ values measured for 150 GeV $\pi^-$ showers developing in lead (**b**). The curves in diagram *a* represent Eq. 7.1. From: Acosta, D. et al. (1992). *Nucl. Instr. and Meth.* **A316**, 184. Experimental data from [3, 4]

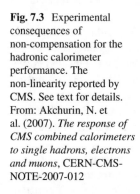

**Fig. 7.3** Experimental consequences of non-compensation for the hadronic calorimeter performance. The non-linearity reported by CMS. See text for details. From: Akchurin, N. et al. (2007). *The response of CMS combined calorimeters to single hadrons, electrons and muons*, CERN-CMS-NOTE-2007-012

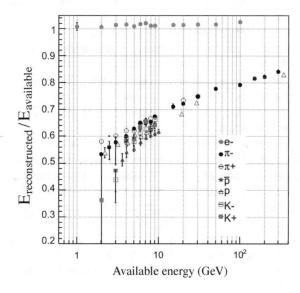

A direct consequence of the energy dependence of $\langle f_{em} \rangle$ is that calorimeters for which $e/h \neq 1$ are by definition *non-linear* for hadron detection, since the response to hadrons is determined by $\langle f_{em} \rangle + \left[1 - \langle f_{em} \rangle\right] h/e$, and thus energy dependent. This is confirmed by many sets of experimental data, for example the ones reported for CMS [6] shown in Fig. 7.3.

Event-to-event fluctuations in $f_{em}$ are large and non-Poissonian [3], as illustrated in Fig. 7.2b. If $e/h \neq 1$, these fluctuations tend to dominate the hadronic energy resolution and their asymmetric characteristics are reflected in the response function [2]. It is often assumed that the effect of non-compensation on the energy resolution is energy independent, and should thus be described i.e., as a "constant term" added in quadrature to the stochastic term:

$$\frac{\sigma}{E} = \frac{a_1}{\sqrt{E}} \oplus a_2 \tag{7.2}$$

where the value of $a_2$ is determined by the degree of non-compensation $(e/h)$. This is incorrect, since it implies that the effect is insignificant at low energies, e.g., 10 GeV, which is by no means the case. The measured effects of *fluctuations* in $f_{em}$ can be described by a term that is very similar to the one used for its energy dependence (7.1). This term should be added in quadrature to the $E^{-1/2}$ scaling term which accounts for all fluctuations that obey Poisson statistics:

$$\frac{\sigma}{E} = \frac{a_1}{\sqrt{E}} \oplus a_2 \left[ \left( \frac{E}{E_0} \right)^{l-1} \right] \tag{7.3}$$

where the parameter $a_2 = |1 - h/e|$ is determined by the degree of non-compensation [7], and $l \sim 0.72$.

The difference between Eqs. 7.2 and 7.3 is illustrated in Fig. 7.4a. The stochastic term has been chosen the same in both cases: $a_1 = 0.5$. For the constant term, we have chosen $a_2 = 0.05$, and the resulting resolution is represented by the dashed curve. The non-compensation parameter in Eq. 7.3 has been given the value $a_2 = 0.3$ and the parameters $E_0$ was given the value 0.7 GeV typical for copper absorber (see Eq. 7.1). The resulting resolution is represented by the solid line. Since the horizontal scale is proportional to $E^{-1/2}$, the stochastic term is represented by a straight (dotted) line through the bottom right corner in Fig. 7.4a.

The difference between the two expressions becomes mainly apparent at energies in excess of 100 GeV, where the non-stochastic term starts to dominate the resolution. However, there are also differences at low energies. Since a constant term contributes very little to the total resolution at low energies, experimental low-energy data should scale approximately with $E^{-1/2}$ if Eq. 7.2 described reality, even in non-compensating calorimeters. On the other hand, in Eq. 7.3 the second term also contributes significantly to the energy resolution at low energies. Therefore, a deviation from $E^{-1/2}$ scaling at low energies constitutes important experimental information.

It turns out that in the energy range covered by the current generation of test beams, i.e., up to 400 GeV, Eq. 7.3 leads to results that are very similar to those obtained with an expression of the type

$$\frac{\sigma}{E} = \frac{c_1}{\sqrt{E}} + c_2 \tag{7.4}$$

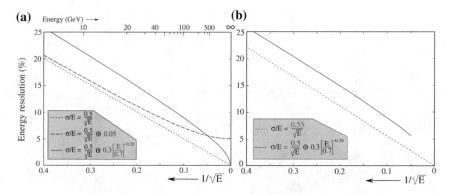

**Fig. 7.4** The effects of non-compensation on the hadronic energy resolution. Comparison of the effects of adding either a constant term or a term as in Eq. 7.3 in quadrature to the stochastic term (**a**). Comparison of Eq. 7.3 (up to 400 GeV) and a stochastic term with a slightly larger coefficient 0.55 instead of 0.50 (**b**) [1]

i.e., a *linear sum* of a stochastic term and a constant term. However this expression suggests that there is complete correlation between the fluctuations that contribute to the two terms, which is nonsense. Figure 7.4b provides the solution of this apparent mystery. The solid line in this figure is exactly the same as in Fig. 7.4a, except that it stops at $E = 400$ GeV, which is the highest pion energy for which experimental data have been reported. This curve runs almost parallel to the dotted line, which represents a stochastic term with a coefficient $a_1 = 0.55$ (in the solid curve, $a_1 = 0.50$). This means that it is practically impossible to distinguish between fits such as

$$\frac{\sigma}{E} = \frac{50\%}{\sqrt{E}} \oplus 30\% \left[ \left( \frac{E}{0.7} \right)^{-0.28} \right]$$

and

$$\frac{\sigma}{E} = \frac{55\%}{\sqrt{E}} + 3.5\%$$

*in the energy range for which experimental data are available.* Experimental data at very high energies would be needed to make that distinction. The observation that experimental resolution data tend to be better described by a linear sum of a stochastic term and a constant term rather than by a quadratic sum of such terms may thus be interpreted as support for Eq. 7.3.

Many sets of experimental hadronic energy resolution data exhibit indeed this characteristic, for example the results reported for ATLAS [8] shown in Fig. 7.5. Following the representation chosen in this book, the energy resolution is plotted on a scale linear in $-E^{-1/2}$. Scaling with $E^{-1/2}$ is thus represented by a straight line through the bottom right corner in this plot. The experimental ATLAS data are located on a line that runs parallel to such a line, indicating that the stochastic term ($c_1$) is ≈80% and the constant term ($c_2$) is ≈5% in this case.

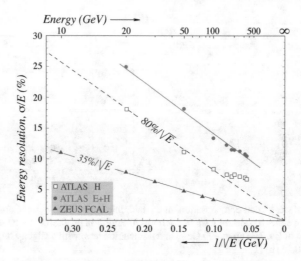

**Fig. 7.5** Experimental consequences of non-compensation for the hadronic calorimeter performance. The energy resolution reported by ATLAS [8], both for the Tilecal in stand-alone mode and for the combination of the em and hadronic calorimeter sections. For comparison, the hadronic energy resolution reported for the compensating ZEUS calorimeter [9] is shown as well. See text for details [1]

Figure 7.5 also shows another interesting phenomenon, namely that the ATLAS hadronic energy resolution was actually measured to be *better* when the hadronic calorimeter section (Tilecal) was used in stand-alone mode, rather than in combination with the LAr ECAL. The reason for this is the fact that these calorimeter sections have different $e/h$ values. For the Tilecal, an $e/h$ value of $1.336 \pm 0013$ has been reported [10], while the value for the Pb/LAr ECAL, which unlike the Tilecal is very insensitive to the neutrons produced in the shower development, is estimated at $\sim 1.5$ [1]. Typically, the energy deposited by showering hadrons and jets is shared between these compartments, and the large event-to-event fluctuations in this energy sharing translate into an additional contribution to the hadronic energy resolution. This contribution is absent when the showers develop entirely in the Tilecal, hence the better energy resolution. However, this better resolution is still considerably worse than that of the compensating ZEUS calorimeter, which is shown for reference purposes in Fig. 7.5. Especially at energies above 100 GeV, the advantages of compensation are very substantial.

The discrepancy between the hadronic energy resolutions measured for the hadronic section alone and for the total calorimeter system is even much larger for the CMS experiment, where the crystal em section has a value of 2.4, while $e/h = 1.3$ for the hadronic section. The hadronic performance of this calorimeter system was systematically studied with various types of particles ($e, \pi, K, p, \bar{p}$), covering a momentum range from $1 - 300$ GeV/$c$. Results are shown in Fig. 7.3. It turned out that the response strongly depends on the starting point of the showers [11]. Figure 7.6 shows results for two event samples, selected on that basis: show-

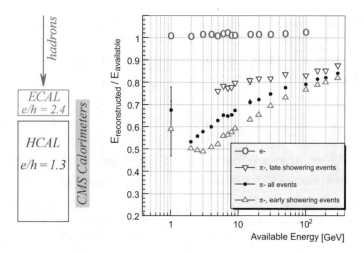

**Fig. 7.6** The response to electrons and pions as a function of energy, for the CMS barrel calorimeter. The pion events are subdivided into two samples according to the starting point of the shower, and the pion response is also shown separately for these two samples. From: Akchurin, N. and Wigmans, R. (2012). *Nucl. Instr. and Meth.* **A666**, 80

ers starting in the em section ($\triangle$) or in the hadronic section ($\triangledown$). At low energies, the response is more than 50% larger for the latter (penetrating) events. In practice in an experiment, it is often hard/impossible to determine where the shower starts, especially if these pions are traveling in close proximity to other jet fragments (e.g., photons from $\pi^0$ decay) which develop showers in the em section.

A different calorimeter response to the em and non-em components of hadron showers also leads to differences in the response functions for different types of hadrons. The absorption of different types of hadrons in a calorimeter may differ in very fundamental ways, as a result of applicable conservation rules. For example, in interactions induced by a proton or neutron, conservation of baryon number has important consequences. The same is true for strangeness conservation in the absorption of kaons. This has implications for the way in which the shower develops. For example, in the first interaction of a proton, the leading particle has to be a baryon. This precludes the production of an energetic $\pi^0$ that carries away most of the proton's energy and thus reduces the energy fraction contained in the em shower component, in comparison with pion-induced showers where no such limitations exist.

In pion-induced showers it is not at all uncommon that most of the energy carried by the incoming particle is transferred to a $\pi^0$. The resulting shower is in that case almost completely electromagnetic. This phenomenon is the reason for the asymmetric distribution seen in Fig. 7.2b. The production of $\pi^0$s in hadron showers is a "one way street" phenomenon: $\pi^0$s may be produced by other hadrons in any stage of the shower development, but they don't produce other hadrons themselves. Therefore, events in which most of the energy carried by the incoming particle is transferred

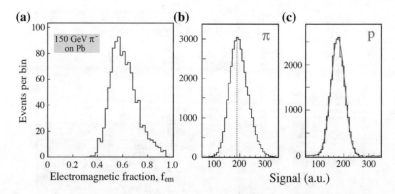

**Fig. 7.7** Event-to-event fluctuations in the em fraction of 150 GeV $\pi^-$ showers in lead (**a**). From: Acosta, D. et al. (1992). *Nucl. Instr. and Meth.* **A316**, 184.Signal distributions for 300 GeV pions (**b**) and protons (**c**) detected with a quartz-fiber calorimeter. The curve represents the result of a Gaussian fit to the proton distribution. From: Akchurin, N. et al. (1998). *Nucl. Instr. and Meth.* **A408**, 380

to a charged pion does not necessarily lead to a small em fraction for the shower as a whole. The limitations on $\pi^0$ production also lead to a smaller value of $\langle f_{em} \rangle$ in baryon induced showers, compared to pion induced ones (Eq. 7.1).

Experimental studies have confirmed these effects [12, 13]. Figure 7.7 shows the signal distributions measured for 300 GeV pions ($b$) and protons ($c$), respectively. The pion distribution resembles that of the fluctuations in the em fraction measured for 150 GeV $\pi^-$ showers developing in lead (Fig. 7.7a). The signal distribution for protons is much more symmetric, as indicated by the Gaussian fit. This is because the em component of proton-induced showers is typically populated by $\pi^0$s that share the energy contained in this component more evenly than in pion-induced showers. The figure also shows that the rms width of the proton signal distribution is significantly smaller (by ~20%) than for the pions. The average signal per GeV deposited energy is smaller for the protons than for the pions, by about 10%. This is also a consequence of the limitations on $\pi^0$ production that affect the proton signals in this non-compensating calorimeter ($e/h > 1$). So while the response to protons is smaller in this calorimeter, the energy resolution is better. This will be true for *all* non-compensating calorimeters, but the extent of the proton–pion differences depends of course on the degree of non-compensation.

Similar effects are expected to play a role for the detection of kaons, where $\pi^0$ production is limited as a result of strangeness conservation in the shower development. Also in this case, the absence of a leading particle effect may be expected to make the fluctuations in $f_{em}$ smaller and more symmetrically distributed than for pions.

## 7.3  The *e/mip* Ratio and Its Effects for Jet Detection

Figure 7.1 shows that the response of typical calorimeters to the em shower component not only differs from that to the non-em component, but it also differs from the response to minimum ionizing particles. The reasons for this are discussed in Sect. 3.5.2. This effect, which I will refer to as $e/mip \neq 1$, may have important consequences for some aspects of the calorimeter performance, even if the calorimeter can be made compensating ($e/h = 1.0$). For example, at low energies, the probability that a hadron is stopped in the calorimeter before it has an opportunity to initiate a nuclear reaction, and thus start a shower, rapidly increases. The entire energy of such hadrons is used to ionize the calorimeter material. There are thus no losses due to invisible energy and, as a result, the calorimeter response to such hadrons is larger than that to hadrons that do develop showers while being absorbed. The response is actually similar to that for muons, which deposit their energy in the same way.

This is illustrated in Fig. 7.8, which shows the response of the $^{238}$U/plastic-scintillator ZEUS calorimeter to low-energy charged hadrons [14]. This calorimeter had an $e/h$ value very close to 1.0, but since the $e/mip$ value was about 0.6 (see Sect. 7.4), the hadronic response increased for energies below a few GeV, reflecting the increasingly *mip*-like absorption process. This calorimeter was thus quite non-linear for hadrons with energies less than 5 GeV, which had important consequences for the performance for jet detection. A jet is a collection of particles (mainly pions and $\gamma$s) produced in the fragmentation of a quark or gluon. Relatively low-energy fragments account for a significant fraction of the energy of high-energy jets, such as the ones produced in the hadronic decay of the $W$ and $Z$ intermediate vector bosons. Figure 7.9 shows the distribution of the energy released by $Z^0$s (decaying through the process $Z^0 \rightarrow u\bar{u}$) and Higgs bosons (decaying into a pair of gluons) at rest that is carried by charged final-state particles with a momentum less than 5 GeV/$c$. The figure shows that, most probably, 21% of the energy equivalence of the $Z^0$ mass is carried by such particles, and the event-to-event fluctuations are such that this frac-

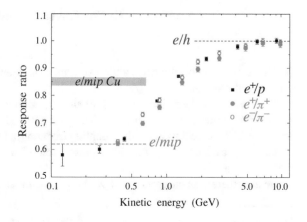

**Fig. 7.8** The ratio of the responses of the (compensating) ZEUS calorimeter to electrons and (low-energy) hadrons. This ratio equals 1.0 for energies above ~10 GeV. At low energies, the hadron response increases because of the absence of nuclear interactions, and the associated losses in nuclear binding energy [1]. Experimental data from [14]

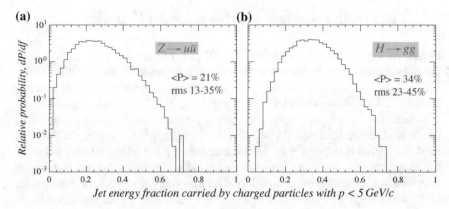

**Fig. 7.9** Distribution of the fraction of the energy released by hadronically decaying $Z^0$ (**a**) and $H^0$ (**b**) bosons at rest that is carried by charged final-state particles with a momentum less than 5 GeV/$c$ [15]

tion varies between 13 and 35% (for a $1\sigma_{rms}$ interval). For Higgs bosons decaying into a pair of gluons, the average fraction is even larger, 34%, with rms variations between 23 and 45%.

As a result of the important contribution from soft jet fragments, and the large event-to-event fluctuations in this contribution, the energy resolution for intermediate vector bosons measured with the compensating ZEUS uranium calorimeter turned out to be worse than expected on the basis of the single-pion resolution.

As shown in the next section, this effect would have been considerably smaller if the calorimeter had used a lower-$Z$ absorber material such as copper. In that case the $e/mip$ value of the calorimeter would have been $\sim$0.85 instead of 0.6 and Fig. 7.8 indicates that the decrease of the response to low-energy hadrons would have been a factor of three smaller than for uranium.

## 7.4  The $e/mip$ and $e/h$ Ratios for Different Types of Calorimeters

In this section, we briefly review the expected $e/mip$ and $e/h$ values for different types of calorimeters.

The $Z$ values of the active and passive calorimeter materials are the most important factors determining the $e/mip$ value. This is illustrated in Fig. 7.10, which shows results of EGS4 calculations of the sampling fraction for (10 GeV)[1] electron showers in sampling calorimeters with either plastic scintillator or liquid argon as active material, as a function of the $Z$ value of the absorber material [16]. The thickness of the absorbers was chosen to be 1 $X_0$ in these simulations, the active layers were

---

[1]These results are more or less independent of the shower energy.

**Fig. 7.10** The $e/mip$ ratio
for sampling calorimeters as
a function of the $Z$ value of
the absorber material, for
calorimeters with plastic
scintillator or liquid argon as
active material. Experimental
data are compared with
results of EGS4 Monte Carlo
simulations. From:
Wigmans, R. (1987). *Nucl.
Instr. and Meth.* **A259**, 389

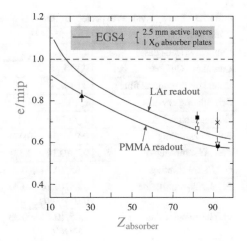

2.5 mm thick. The $e/mip$ ratio, which follows directly from these results, gradually
decreases when the $Z$ value of the absorber increases. The $e/mip$ values are also
systematically larger when liquid argon ($Z = 18$) readout is used instead of plastic
scintillator. It is the *difference* in $Z$ values between active and passive media that
determines the $e/mip$ ratio. The simulations show that this ratio can even be made
larger than 1.0 if $Z_{active} > Z_{passive}$, e.g., in Al/Lar calorimeters.

Experimental results on the value of $e/mip$ are included in this figure. They
concern measurements on calorimeters based on iron, lead or uranium as absorber
material. The results are in reasonable agreement with the simulations.

The $e/h$ values are based on the responses to the various components of hadron
showers and the fractions of their contribution to the shower development. All
responses are normalized to the response to mips, so that

$$\frac{e}{h} = \frac{e/mip}{f_{rel} \cdot rel/mip \; + \; f_p \cdot p/mip \; + \; f_n \cdot n/mip} \tag{7.5}$$

where the non-em component consists of relativistic particles, protons and neutrons,
which carry fractions $f_{rel}$, $f_p$ and $f_n$ of the non-em energy, respectively. The expected
values are summarized in Table 7.1. The $e/mip$ values for the calorimeters in this
Table are taken from Fig. 7.10. The fractions of the non-em shower component carried
by relativistic particles and by protons and neutrons from nuclear breakup were taken
from Monte Carlo simulations by Gabriel [17]. It was assumed that signal quenching
by densely ionizing spallation protons does not play a role in LAr, and that evaporation
neutrons do not generate signals in this medium. For more background information
on the underlying assumptions that were used we refer to [1].

**Table 7.1** The $e/h$ values expected for different types of calorimeters

| Calorimeter type | Expected $e/h$ | Value (range) |
|---|---|---|
| Čerenkov (lead-glass) | $1/0.14$ | $\approx 7$ |
| Čerenkov (Cu/quartz fiber) | $0.8/0.14$ | $5\text{--}6$ |
| Čerenkov (Pb/quartz fiber) | $0.6/0.14$ | $\approx 4$ |
| Scintillating crystals | $1/(0.14 + 0.36\,p/mip)$ | $>2$ |
| Homogeneous LAr | $1(/0.14 + 0.42)$ | $\approx 1.8$ |
| Fe/LAr sampling | $0.9/(0.14 + 0.42)$ | $\approx 1.6$ |
| Pb/LAr sampling | $0.7/(0.14 + 0.33)$ | $\approx 1.5$ |
| $^{238}$U/Lar sampling | $0.65/(0.14 + 0.33 \times 1.2)$ | $1.1\text{--}1.2$ |
| Fe/plastic-scintillator | $0.85/(0.14 + 0.42\,p/mip + 0.05\,n/mip)$ | $\leq 1.5$ |
| Pb/plastic-scintillator | $0.65/(0.14 + 0.33\,p/mip + 0.09\,n/mip)$ | $\leq 1.3$ |
| $^{238}$U/plastic-scintillator | $0.6/(0.14 + 0.33\,p/mip + 0.12\,n/mip)$ | $\leq 1.2$ |

## 7.5 Longitudinal Segmentation of Non-compensating Calorimeters

In Sect. 3.5.2, it was shown that the sampling fraction of em showers decreases as the shower develops. Depending on the specifics of the calorimeter structure, that effect may be as large as 25% (Fig. 3.13). As shown in Chap. 9, this causes major problems for the calibration of longitudinally segmented em calorimeters.

The problems are even much larger for hadron calorimeters, especially when the different longitudinal calorimeter segments have different $e/h$ values, as in CMS (Fig. 7.6). A common misconception is that the em shower fraction in the development of hadron showers is limited to the em section of the calorimeter system and that it therefore is OK to have longitudinal sections with different $e/h$ values. However, Fig. 3.6 demonstrates that $\pi^0$ production may occur anywhere in the development of hadron showers, and is by no means limited to the early stages. Therefore, a calorimeter that consists of longitudinal segments with different $e/h$ values will be subject to *additional* signal fluctuations for the detection of hadrons, compared to a calorimeter of which all segments have the same $e/h$ values.

## 7.6 Summary

The problems that limit the performance of hadron calorimeters can be summarized as follows.

- In the absorption of high-energy hadrons, some fraction of the energy is essentially *invisible*, it cannot contribute to the calorimeter signals.

- This fraction fluctuates wildly from one event to the next. These fluctuations are non-Poissonian.
- This leads to hadronic signal non-linearity, non-Gaussian response functions and poor hadronic energy resolutions.
- The response function of the calorimeter depends on the type of hadron.
- Dividing a calorimeter into longitudinal segments with different $e/h$ values makes these problems worse.
- The effects may be worse for jets than for single hadrons.

In the next chapter, the methods that have been developed to mitigate/solve these problems are described.

# References

1. Wigmans, R.: Calorimetry—Energy Measurement in Particle Physics, 2nd edn. International Series of Monographs on Physics, vol. 168. Oxford University Press, Oxford (2017)
2. Livan, M., Wigmans, R.: Misconceptions about calorimetry. Instruments **1**, 3 (2017). https://doi.org/10.3390/instruments1010003
3. Acosta, D., et al.: Nucl. Instrum. Methods **A316**, 184 (1992)
4. Akchurin, N., et al.: Nucl. Instrum. Methods **A399**, 202 (1997)
5. Gabriel, T.A., et al.: Nucl. Instrum. Methods **A338**, 336 (1994)
6. Akchurin, N., et al.: The response of CMS combined calorimeters to single hadrons, electrons and muons, CERN-CMS-NOTE-2007-012 (2007)
7. Groom, D.E.: Nucl. Instrum. Methods **A572**, 633 (2007)
8. Abat, E., et al.: Nucl. Instrum. Methods **A621**, 134 (2010)
9. Behrens, U., et al.: Nucl. Instrum. Methods **A289**, 115 (1990)
10. Adragna, P., et al.: Nucl. Instrum. Methods **A606**, 362 (2009)
11. Akchurin, N., Wigmans, R.: Nucl. Instrum. Methods **A666**, 80 (2012)
12. Akchurin, N., et al.: Nucl. Instrum. Methods **A408**, 380 (1998)
13. Adragna, P., et al.: Nucl. Instrum. Methods **A615**, 158 (2010)
14. Andresen, A., et al.: Nucl. Instrum. Methods **A290**, 95 (1990)
15. Webber, B.: Private Communication. Cambridge University, Cambridge (2016)
16. Wigmans, R.: Nucl. Instrum. Methods **A259**, 389 (1987)
17. Gabriel, T.A.: Proceedings of the Workshop on Compensated Calorimetry, Pasadena, Internal Report CALTECH–68–1305, p. 238 (1985)

# Chapter 8
# Methods to Improve Hadronic Calorimeter Performance

## 8.1 Introduction

The root cause of the poor performance of hadron calorimeters is thus the invisible energy. Because some fraction of the energy carried by the hadrons and released in the absorption process does not contribute to the signal, the response to the non-em shower component is typically smaller than that to the em shower component. And the characteristic features of the energy sharing between these two components lead to hadronic signal non-linearity, a poor energy resolution and a non-Gaussian response function.

To mitigate these effects, one has thus to use a measurable quantity that is correlated to the invisible energy. The stronger that correlation, the better the hadronic calorimeter performance may become. In this chapter, two such measurable quantities are discussed: the kinetic energy released by neutrons in the absorption process (Sect. 8.2) and the total non-em energy (Sect. 8.3). In Sect. 8.4, the beneficial effects of both methods are compared.

## 8.2 Compensation

The first successful attempt to mitigate the effects described in the previous section involved a calorimeter that used depleted uranium as absorber material [1]. The underlying idea was that the fission energy released in the absorption process would compensate for the invisible energy losses. By boosting the non-em calorimeter response ($h$) this way, the $e/h$ ratio would decrease and, as a matter of good fortune, reach the (ideal) value of 1.0. This is the reason why calorimeters with $e/h = 1.0$ have become known as *compensating* calorimeters. However, it turned out that nuclear fission was neither essential nor sufficient to reach the compensation condition. Several uranium calorimeters that were built after this initial success were found to have $e/h$ values that were larger [2], and in one case significantly smaller than 1.0 [3].

© Springer Nature Switzerland AG 2019
M. Livan and R. Wigmans, *Calorimetry for Collider Physics, an Introduction*,
UNITEXT for Physics, https://doi.org/10.1007/978-3-030-23653-3_8

On the other hand, it was demonstrated that the compensation condition can also be obtained in calorimeters that use lead as absorber material [4].

In order to understand how compensation can be achieved, one should understand in detail the response to the various types of particles that contribute to the calorimeter signals. Let us look at the generic definition of the $e/h$ value (Eq. 7.5). Because of the invisible energy, the "natural" value is larger than 1.0, and quite a bit larger if the calorimeter only responds to relativistic shower particles (which is the case for Čerenkov calorimeters). In order to bring the $e/h$ value down, two terms are very important: the numerator ($e/mip$) and the $n/mip$ term in the denominator. As shown in the previous chapter (Fig. 7.10), the $e/mip$ value can be significantly smaller than 1.0 when high-$Z$ absorber materials are used. This was definitely an important factor contributing to the initial success of the uranium based calorimeter, where $e/mip \approx 0.6$.

However, most important in this context are the neutrons. Neutrons carry typically not more than $\sim 10\%$ of the non-em shower energy. However, their contribution to the calorimeter signals may be much larger than that. This is because neutrons only lose their energy through the products of the nuclear reactions they undergo. Most prominent at the low energies typical for hadronic shower neutrons is *elastic scattering*. As shown in Sect. 3.5.2 (Table 3.1), MeV-type neutrons sent into a $Pb/H_2$ structure (50/50 in terms of numbers of nuclei), transfer 98% of their kinetic energy to hydrogen nuclei, and only 2% to lead. Since the sampling fraction for charged particles (mips) amounts to 2.2% in this structure, the potential for *signal amplification through neutron detection* (SAND) is enormous, especially when the recoil protons produced in the active material directly contribute to the calorimeter signal.

Hydrogenous active material is an extremely efficient medium for SAND in calorimeters. Nowhere has the role of hydrogen been demonstrated more dramatically than in the L3 uranium/gas calorimeter [5]. Figure 8.1a shows the signals of this calorimeter for pions and for electrons, as a function of energy, for two different gas mixtures: Argon/$CO_2$ and isobutane. For the electron signals, the choice of gas made no significant difference. However, the pion response doubled when isobutane ($C_4H_{10}$) was used instead of argon/$CO_2$. The L3 group also tested other gas mixtures. It turned out that by changing the hydrogen content of the gas mixture used in the wire chambers that produced the calorimeter signals, the $\pi/e$ response ratio could be changed by as much as a factor of two. By choosing the proper mixture, the responses to em and hadronic showers could be equalized (Fig. 8.1b).

Compensation can also be achieved in other types of calorimeters, provided that the active material contains hydrogen. Plastic scintillator is well suited, since the recoil protons contribute to the calorimeter signals. Because of signal quenching (Fig. 4.3), the $n/mip$ ratio (and thus the $e/h$ value) is much less sensitive to the precise amount of hydrogen than indicated in Table 3.2 and measured by L3. By carefully choosing the relative amount of hydrogen in the calorimeter structure, compensation can be achieved. This has been demonstrated experimentally for plastic-scintillator structures with Pb or $^{238}U$ as absorber material (Fig. 8.2).

All compensating calorimeters rely on the contribution of neutrons to the hadronic signals. By properly amplifying the neutron signals (with respect to those from

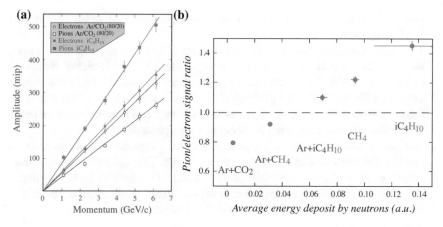

**Fig. 8.1** Signals for pion and electron showers in the L3 uranium/gas calorimeter, for 2 different gas mixtures in the readout chambers (**a**). Pion/electron response ratio as a function of the hydrogen content of the gas mixture (**b**). From: Galaktionov, Y. et al. (1986). *Nucl. Instr. and Meth.* **A251**, 258

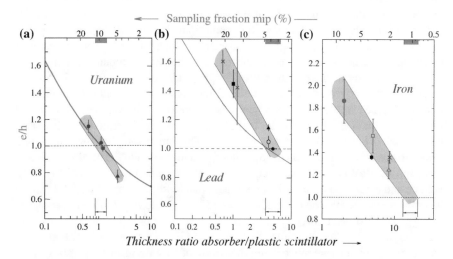

**Fig. 8.2** Experimental data on the $e/h$ values measured for calorimeters based on plastic scintillator as active material and using either $^{238}$U (**a**), lead (**b**) or iron (**c**) as absorber material. The $e/h$ values are plotted as a function of the ratio of the thicknesses of the absorber and scintillator plates (bottom scale), i.e., as a function of the sampling fraction for mips (top scale). The curves represent the results of simulations based on a neutron transportation Monte Carlo program [6]

charged shower particles depositing the same amount of energy), one can *compensate* for the invisible-energy losses. Therefore, the essential ingredients for a compensating calorimeter are:

- One needs to have a *sampling* calorimeter. Compensation can never be achieved in a homogeneous one.
- The active material needs to contain *hydrogen* and be sensitive to the signals from recoil protons produced by elastic neutron scattering. This is nicely visualized in Fig. 11.8, which shows typical time structures of electron and pion signals detected with the SPACAL calorimeter. The hadron signals have a "late tail" with a time constant of $\sim$10 ns, which is caused by slow recoil protons produced in the absorption process. Such a tail is absent in the electron signals.
- The calorimeter needs to have a precisely tuned *sampling fraction*, in order to amplify the neutron signals by the proper factor. This optimal sampling fraction is $\sim$10% for U/plastic-scintillator (Fig. 8.2a), $\sim$3% for Pb/plastic-scintillator (Fig. 8.2b). Available experimental data suggest that it might even be possible to achieve compensation in Fe/plastic scintillator devices (Fig. 8.2c).

Indeed, it turned out that the mentioned effects of non-compensation on energy resolution, linearity and line shape, as well as the associated calibration problems [7], are absent in compensating calorimeters, except at energies that are so low that nuclear reactions cease to play a role (Fig. 7.8). However, it also turned out that fission had nothing to do with this, and that the use of uranium was neither necessary nor sufficient for reaching the compensation condition. The crucial element was rather the active material of the sampling calorimeter, which has to be very efficient in detecting the numerous neutrons produced in the shower development process.

Compensation can thus be achieved in sampling calorimeters with high-$Z$ absorber material and hydrogenous active material. It requires a very specific sampling fraction, so that the response to shower neutrons is boosted by the precise factor needed to equalize $e$ and $h$. For example, in Pb/scintillating-plastic structures, this sampling fraction is $\sim$3% for showers [4, 8, 9]. This small sampling fraction sets a lower limit on the contribution of sampling fluctuations, while the need for efficient detection of MeV-type neutrons requires signal integration over a relatively large volume and at least 30 ns. Yet, the experiment that holds the current world record in hadronic energy resolution (ZEUS, $\sigma/E \sim 35\%/\sqrt{E}$, see Fig. 7.5) used a compensating $^{238}$U calorimeter [10]). The hadronic signal distributions for the compensating Pb/scintillator SPACAL calorimeter are shown in Fig. 8.3. Especially at high energies, these resolutions are much better than those of the calorimeters operating in the LHC experiments.

In compensating calorimeters, the total kinetic energy of the neutrons produced in the hadronic shower development thus represents the measurable quantity that is correlated to the invisible energy. As a matter of fact, this correlation is better for lead than for uranium absorber, since the neutrons produced in fission are not correlated to the nuclear binding energy loss. This was experimentally confirmed by ZEUS, who measured the contributions from *intrinsic* fluctuations (i.e., a measure for the degree of correlation mentioned here) to the hadronic energy resolution and found it to be better for lead than for uranium: $13.4\%/\sqrt{E}$ versus $20.4\%/\sqrt{E}$ (Table 6.2). The relative magnitude of the signal provided by the neutrons can be tuned to achieve equality of the electromagnetic and non-electromagnetic calorimeter

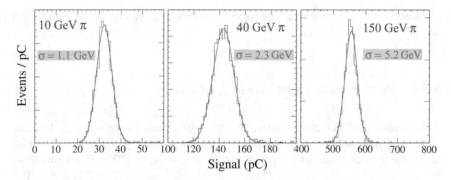

**Fig. 8.3** Signal distributions for hadrons of 10, 40 and 150 GeV measured with SPACAL. From: Acosta, D. et al. (1991). *Nucl. Instr. and Meth.* **A308**, 481

responses ($e/h = 1.0$), by means of the sampling fraction. This mechanism works because the calorimeter response to charged shower particles is much more sensitive to a change in the sampling fraction than the response to neutrons.

## 8.3 Dual-Readout Calorimetry

The dual-readout approach [11] aims to achieve the advantages of compensation without the disadvantages mentioned in the previous section:

- The need for high-$Z$ absorber material, and the associated small $e/mip$ value, which causes non-linearities at low energy and deteriorates the jet performance.
- A small sampling fraction, which limits the em energy resolution.
- The need to detect MeV-type neutrons efficiently, which implies integrating the signals over relatively large detector volumes and long times.

The purpose of the dual-readout technique is to measure the em shower fraction ($f_{em}$) *event by event*. If successful, this would make it possible to diminish/eliminate the effects of fluctuations in $f_{em}$ on the hadronic calorimeter performance. This was in itself not a new idea. Starting around 1980, attempts have been made to disentangle the energy deposit profiles of hadronic showers with the goal to identify the em components, which are typically characterized by a high localized energy deposit [12]. Such methods are indeed rather successful for isolated high-energy hadron showers, but fail at low energy, and in particular when a number of particles develop showers in the same vicinity, as is typically the case for jets.

The dual-readout method exploits the fact that the energy carried by the non-em shower components is mostly deposited by non-relativistic shower particles (protons), and therefore does not contribute to the signals of a Čerenkov calorimeter. By measuring simultaneously $dE/dx$ and the Čerenkov light generated in the shower absorption process, it is possible to determine $f_{em}$ event by event and thus eliminate (the effects of) its fluctuations. The correct hadron energy can be determined from a combination of both signals.

## 8.3.1  Dual-Readout Analysis Procedures

A dual-readout calorimeter produces two types of signals for the showers developing in it, a scintillation signal ($S$) and a Čerenkov signal ($C$). Both signals can be calibrated with electrons of known energy $E$, so that $\langle S \rangle = \langle C \rangle = E$ for em showers, and the calorimeter response to em showers, $R_{em} = \langle S \rangle / E = \langle C \rangle / E = 1$. For a given event, the hadronic signals of this calorimeter can then be written as

$$S = E\left[ f_{em} + \frac{1}{(e/h)_S}(1 - f_{em}) \right]$$

$$C = E\left[ f_{em} + \frac{1}{(e/h)_C}(1 - f_{em}) \right] \tag{8.1}$$

i.e., as the sum of an em shower component ($f_{em}$) and a non-em shower component ($1 - f_{em}$). The contribution of the latter component to the reconstructed energy is weighted by a factor $h/e$. When $f_{em} = 1$ or $e/h = 1$, the hadronic shower response is thus the same as for electrons: $R = 1$. However, in general $f_{em} < 1$ and $e/h \neq 1$, and therefore the hadronic response is different from 1. The reconstructed energy is thus different (typically smaller) than $E$.

The dual-readout method works thanks to the fact that $(e/h)_S \neq (e/h)_C$. The larger the difference between both values, the better. The em shower fraction $f_{em}$ and the shower energy $E$ can be found by solving Eqs. 8.1, using the measured values of the scintillation and Čerenkov signals and the *known e/h* ratios of the Čerenkov and scintillator calorimeter structures.

Looking at Eqs. 8.1, we see that the ratio of the two measured signals $S$ and $C$ is *independent of the shower energy E*. There is thus a one-to-one correspondence between this measured signal ratio and the value of the em shower fraction, $f_{em}$. This fraction can thus be determined for each individual event, and therefore the effects of fluctuations in $f_{em}$ can be eliminated. Just as in compensating calorimeters, where these fluctuations are eliminated by design, this is the most essential ingredient for improving the quality of hadron calorimetry.

The merits of this principle were first demonstrated by the DREAM Collaboration [13], with a Cu/fiber calorimeter. Scintillating fibers measured $dE/dx$, quartz fibers the Čerenkov light. The response ratio of these two signals was related to $f_{em}$ as

$$\frac{C}{S} = \frac{f_{em} + 0.21\,(1 - f_{em})}{f_{em} + 0.77\,(1 - f_{em})} \tag{8.2}$$

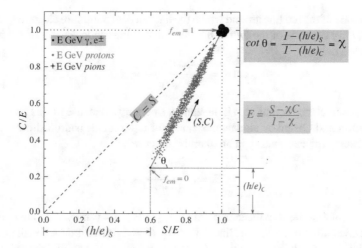

**Fig. 8.4** Graphic representation of Eqs. 8.3 [11, 14]. The data points for hadron showers detected with a dual-readout calorimeter are located around the straight (red) line in this diagram. The data points for em showers in this calorimeter are clustered around the point where this line intersects the $C = S$ line, i.e., the point (1,1). From: Lee, S., Livan, M. and Wigmans, R. (2018). *Rev. Mod. Phys.* **90**, 025002

where 0.21 and 0.77 represented the $h/e$ ratios of the Čerenkov and scintillator calorimeter structures, respectively. They used the measured data to show that the broad, asymmetric signal distribution typical for a non-compensating calorimeter is in fact a superposition of many individual, narrow, Gaussian distributions for events with the same $f_{em}$ but a different central value that gradually increased with $f_{em}$. The overall signal distribution just reflected the distribution of $f_{em}$ values in the events. These results are shown in Fig. 8.11a, b and it is instructive to compare them with the $f_{em}$ distribution from Fig. 7.7a.

Let us now look again at Eqs. 8.1, and rewrite these as

$$S/E = (h/e)_S + f_{em}[1 - (h/e)_S]$$
$$C/E = (h/e)_C + f_{em}[1 - (h/e)_C] \tag{8.3}$$

Figure 8.4 shows that the experimental data points for hadron showers detected with a dual-readout calorimeter are thus located around a straight (red) line in the $C/E$ versus $S/E$ diagram. This line links the point $[(h/e)_S, (h/e)_C]$, for which $f_{em} = 0$, with the point (1,1), for which $f_{em} = 1$. The experimental data points for electron showers are concentrated around the latter location in this diagram, where the red line intersects with the $C = S$ line. By projecting these data points on the horizontal and vertical axes, the raw scintillation and Čerenkov signal distributions are obtained, and the shape of these distributions is indicative for the $f_{em}$ distribution in the event sample. Different $f_{em}$ distributions will thus lead to different signal distributions, as in Fig. 7.7b, c.

The slope of the red line around which the hadron data points are clustered, i.e., the angle $\theta$, only depends of the two $e/h$ values, and is thus *independent of the hadron energy*. We define

$$\cot \theta = \frac{1 - (h/e)_S}{1 - (h/e)_C} = \chi \tag{8.4}$$

and the parameter $\chi$ is thus also independent of energy. Because of this feature, the scintillation and Čerenkov signals measured for a particular hadron shower can be used to reconstruct its energy in an unambiguous way:

$$E = \frac{S - \chi C}{1 - \chi} \tag{8.5}$$

This equation implies that the data point $(S, C)$ in Fig. 8.4 is moved up along the red straight line until it intersects the line defined by $C = S$. If this is done for all hadronic data points, the result is a collection of data points that cluster around the point $(1,1)$, just like the data points for electron showers. Projecting these data points on the axes of this diagram gives the scintillation $(x)$ and Čerenkov $(y)$ signal distributions.

The dual-readout procedure thus effectively uses the measured signals to determine the em shower fraction, $f_{em}$, and then calculates what the signals would be if $f_{em}$ was 1.0, i.e., as if the shower was entirely electromagnetic. The actual $f_{em}$ distribution for showers produced in the absorption of a sample of hadrons of the same type and energy is therefore not a factor that affects the energy measurement for that event sample. A dual-readout calorimeter is therefore *linear* for hadron detection, since the correct energy is reproduced in each case.

Interestingly, a dual-readout calorimeter will also produce signal distributions with the same average value for event samples of pions, protons and kaons of the same energy. The $f_{em}$ distributions are quite different for showers produced by these different types of hadrons, as a result of conservation of baryon number and strangeness in the shower development. This prevents the production of a very energetic, leading $\pi^0$ in the case of protons and kaons, respectively. These differences are illustrated in Fig. 8.4, where the protons and pions are represented by red and blue data points, respectively. Measurements with conventional calorimeters have clearly shown significant differences between the response functions of protons and pions (Fig. 7.7). Figure 8.5c illustrates that the mentioned effects do not play a role for dual-readout calorimeters. The relationship (8.5) is universally valid for all types of hadrons, and also for jets.

## 8.3.2 Some Experimental Results

Figure 8.5 shows an example of results obtained in practice with a dual-readout fiber calorimeter built and tested as part of the RD52 detector R&D program [15]. The fact that $\theta$ and $\chi$ are independent of the energy and the particle type offers the possibility

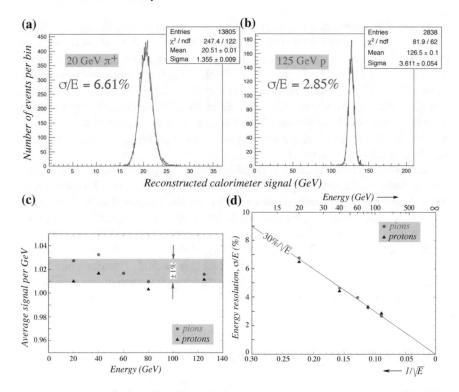

**Fig. 8.5** Results obtained with the RD52 lead-fiber dual-readout calorimeter. Shown are the measured signal distributions for 20 GeV $\pi^+$ (**a**) and 125 GeV protons (**b**), the average signal per unit deposited energy as a function of energy, separately for pions and protons (**c**), and the measured signal energy resolution for protons and pions as a function of energy (**d**). From: Lee, S. et al. (2017). *Nucl. Instr. and Meth.* **A866**, 76

to measure the hadronic energy with unprecedented precision, in test beams at the CERN SPS. It should be emphasized that no information on the beam energy or the hadron type was used in the analyses of the experimental data.

Figure 8.5a, b show signal distributions for 20 GeV pions and 125 GeV protons, respectively. Both distributions are well described by a Gaussian function with a central value of 20.5 GeV and 126.5 GeV, respectively. The relative width, $\sigma/E$, was measured to be 6.61% for the pions and 2.85% for the protons. This corresponds to $29.6\%/\sqrt{E}$ and $31.9\%/\sqrt{E}$, respectively. The narrowness of these distributions reflects the clustering of the data points around the axis of the locus in Fig. 8.4. It should be re-emphasized that the energy of the beam particles was *not* used to obtain these signal distribution. The straight line that was used to fit the experimental data points in the scatter plot (Fig. 8.4) intersected the $C = S$ line at approximately the correct energy. The same was true for particles of other energies, for different types of hadrons and also for multiparticle events, always using the same procedure and the same value for the angle $\theta$.

Figure 8.5c shows that this dual-readout calorimeter is very linear. It produces the same response for pions and protons, within $\approx \pm 1\%$, and, most importantly, the hadron energy was correctly reproduced in this device, of which the energy scale was set with electron showers. Finally, the hadronic energy resolution was measured to scale with $\sim 30\%/\sqrt{E}$ over the full energy range for which the calorimeter was tested, without significant deviations (Fig. 8.5d). Since the calorimeter (including its system of leakage detectors) had a mass of less than 2 tonnes, the effects of fluctuations in (mainly lateral) shower leakage were probably not at all negligible, and even better resolutions may be expected for larger instruments of this type.

In dual-readout calorimeters, the total non-em energy, which can be derived from the measured total energy (Eq. 8.5) and the em shower fraction (Eq. 8.1), thus represents the measurable quantity correlated to the invisible energy. The limitations that apply for compensation do not apply in this case. Any absorber material may be used, as a matter of fact the dual-readout method may even be applied for *homogeneous* calorimeters, such as BGO crystals [16]. The sampling fraction is not restricted and neutron detection is not a crucial ingredient for this method. Therefore, one is considerably less constrained when designing a calorimeter system of this type than in the case of a system based on compensation.

## 8.4   Dual-Readout Versus Compensation

Compensating calorimeters and dual-readout calorimeters both try to eliminate/mitigate the effects of fluctuations in the invisible energy on the signal distributions by means of a measurable variable that is correlated to the invisible energy. As mentioned in the previous subsections, the variables used for this purpose are different in compensating and dual-readout calorimeters. However, with both methods a very significant improvement of the hadronic calorimeter performance is obtained, compared to the standard non-compensating calorimeters used in the current generation of particle physics experiments: the hadronic response is independent of the energy (i.e., the calorimeter is linear) and the type of hadron, the hadronic response function is Gaussian, the hadronic energy resolution is much better and, most importantly, a calibration with electrons also provides the correct energy for hadronic showers.

In Fig. 8.6, the energy resolutions obtained with the best compensating calorimeters, ZEUS [10] and SPACAL [8], are compared with the results obtained with the RD52 dual-readout fiber calorimeter. Figure 8.6b shows that the hadronic RD52 values are actually better than the ones reported by ZEUS and SPACAL, while Fig. 8.6a shows that the RD52 em energy resolution is certainly not worse.

In making this comparison, it should be kept in mind that

1. The em energy resolutions shown for RD52 were obtained with the calorimeter oriented at a much smaller angle with the beam line ($\theta, \phi = 1°, 1.5°$) than the ones for SPACAL ($\theta, \phi = 2°, 3°$) [17]. It has been shown that the em energy

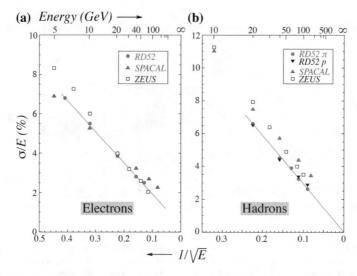

**Fig. 8.6** Energy resolutions reported for the detection of electrons(*a*) and hadrons (*b*) by RD52 [15, 17], SPACAL [8] and ZEUS [10]. From: Lee, S., Livan, M. and Wigmans, R. (2018). *Nucl. Instr. and Meth.* **A882**, 148

resolution is extremely sensitive to the angle between the beam particles and the fiber axis when this angle is very small [18].

2. The instrumented volume of the RD52 calorimeter (including the leakage counters) was less than 2 tonnes, while both SPACAL and ZEUS obtained the reported results with detectors that were sufficiently large (>20 tonnes) to contain the showers at the 99+% level. The hadronic resolutions shown for RD52 are definitely affected by fluctuations in lateral shower leakage, and a larger instrument of this type is thus very likely to further improve the results.

The comparison of the hadron results (Fig. 8.6b) seems to indicate that the dual-readout approach offers better opportunities to achieve superior hadronic performance than compensation. Apparently, in hadronic shower development the correlation with the total nuclear binding energy loss is thus stronger for the total non-em energy (derived from the em shower fraction) than for the total kinetic neutron energy. Intuitively, this is not a surprise, since the total non-em energy consists of other components than just neutrons, and the total kinetic energy of the neutrons is not an exact measure for the *number* of neutrons (which is the parameter expected to be correlated to the binding energy loss).

In order to investigate the validity of this interpretation of the experimental results, Lee et al. performed Monte Carlo simulations of shower development in a block of matter that was sufficiently large to make the effects of shower leakage insignificantly small. Large blocks of copper or lead were used for this purpose [19]. The simulations were carried out with the GEANT4 Monte Carlo package [20] for pions of 10, 20,

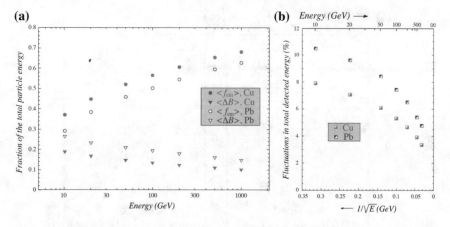

**Fig. 8.7** The average value of the em shower fraction and the average fraction of the energy represented by nuclear binding energy losses, for pions absorbed in large blocks of copper and lead, as a function of the pion energy (**a**). Event-to-event fluctuations in the nuclear binding energy losses, expressed as a fraction of the total detected energy (**b**). From: Wigmans, R. (2018). J. Progr. Part. Nucl. Phys. **103**, 109

50, 100, 200, 500 and 1000 GeV, using the default physics list used in simulations for the CMS and ATLAS experiments at CERN's Large Hadron Collider [21].

Some results of these simulations are shown in Figs. 8.7, 8.8, 8.9 and 8.10. Figure 8.7a shows the average value of the em shower fraction and the average fraction of the pion energy represented by nuclear binding energy losses as a function of the pion energy. There are clear differences between copper and lead. The em shower fraction increases with energy and is larger for copper (in agreement with the measurement results shown in Fig. 7.2a), while the average binding energy losses decrease with energy and are larger for lead. The event-to-event fluctuations in the binding energy loss are also larger for lead, as illustrated in Fig. 8.7b. This figure represents the ultimate precision with which the energy of the pions can be measured in a copper- or lead-based calorimeter in which no effort is made to mitigate the effects of these fluctuations on the hadronic energy resolution.

However, both the non-em energy (which follows directly from $f_{em}$) and the total kinetic neutron energy turned out to be clearly correlated with the nuclear binding energy loss (Fig. 8.8). To examine the degree of correlation, event-by-event ratios were determined. Histograms of these ratios are shown in Fig. 8.9 for 50 GeV $\pi^-$ showers in copper.

These figures confirm that the correlation between the total non-em energy and the nuclear binding energy loss is better than the correlation between the total kinetic neutron energy and the nuclear binding energy loss. This was found to be true both for copper and for lead. These simulations were also used to estimate the effects of the correlations discussed above on the energy resolution for hadron calorimeters that are based on dual-readout or compensation. The results are summarized in Fig. 8.10, for pions in copper (*a*) and lead (*b*). These resolutions should be considered Monte

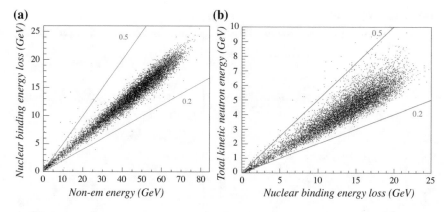

**Fig. 8.8** Scatter plots in which the nuclear binding energy losses for 100 GeV pions absorbed in copper are compared event by event to the non-em energy measured with the dual-readout method (**a**) and the total kinetic energy of the neutrons produced in the absorption process, which is the essential ingredient of compensating calorimeters (**b**). The straight lines represent constant values (0.2 or 0.5) of the ratio between the parameters plotted on the vertical and horizontal axes. Results of GEANT4 Monte Carlo simulations. From: Lee, S., Livan, M. and Wigmans, R. (2018). *Nucl. Instr. and Meth.* **A882**, 148

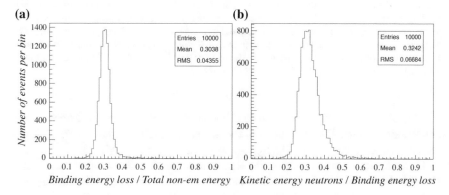

**Fig. 8.9** Distributions of the ratio of the non-em energy and the nuclear binding energy loss (**a**) and the ratio of the total kinetic energy carried by neutrons and the nuclear binding energy loss (**b**) for hadron showers generated by 50 GeV $\pi^-$ in a massive block of copper. Results from GEANT Monte Carlo simulations. From: Lee, S., Livan, M. and Wigmans, R. (2018). *Nucl. Instr. and Meth.* **A882**, 148

Carlo predictions for the *ultimate hadronic energy resolution* that can be achieved with calorimeters using either dual-readout or compensation as the method to mitigate the effects of (fluctuations in) invisible energy. A comparison of these results with those from Fig. 8.7b shows to what extent these methods are successful in that respect, especially at increasing energy.

The resolution limits scale remarkably well with $E^{-1/2}$, in the energy range considered here (10–1000 GeV). Both for lead and for copper absorber, the limits are

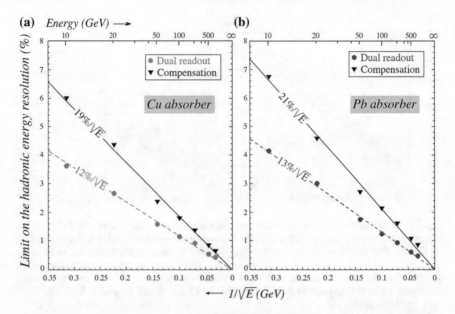

**Fig. 8.10** The limit on the hadronic energy resolution derived from the correlation between nuclear binding energy losses and the parameters measured in dual-readout or compensating calorimeters, as a function of the particle energy. The straight lines represent resolutions of $20\%/\sqrt{E}$ and $10\%/\sqrt{E}$, respectively, and are intended for reference purposes. Results from GEANT Monte Carlo simulations of pion showers developing in a massive block of copper (**a**) or lead (**b**). From: Lee, S., Livan, M. and Wigmans, R. (2018). *Nucl. Instr. and Meth.* **A882**, 148

considerably better for dual-readout calorimeters than for compensating ones, in this entire energy range. The ultimate limit for the energy resolution that can be achieved with calorimetric detection of hadron showers seems to be $\sim 12\%/\sqrt{E}$.

Experimental data obtained by the RD52 Collaboration also support the conclusion that the correlation exploited in dual-readout calorimeters provides a more accurate measurement of the invisible energy. Figure 8.11a, b show that the (Čerenkov) signal from the DREAM fiber calorimeter is actually a superposition of many rather narrow, Gaussian signal distributions. Each sample in Fig. 8.11b contains events with (approximately) the same $f_{em}$ value, i.e., with the same total non-em energy. The dual-readout method combines all these different subsamples and centers them around the correct energy value. The result is a relatively narrow, Gaussian signal distribution with the same central value as for electrons of the same energy.

Figure 8.11d shows that the DREAM (Čerenkov) signal is also a superposition of Gaussian signal distributions of a different type. In this case, each sample consists of events with (approximately) the same total kinetic neutron energy. The dual-readout method may combine all these different subsamples in the same way as described above. In doing so, the role of the total non-em energy is taken over by the total kinetic neutron energy, and the method becomes thus very similar to the one used in compensating calorimeters.

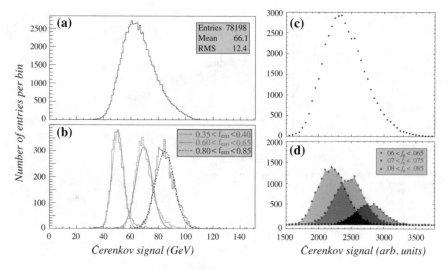

**Fig. 8.11** Distribution of the total Čerenkov signal for 100 GeV $\pi^-$ (**a**) and the distributions for three subsets of events selected on the basis of the electromagnetic shower fraction (**b**). From: Akchurin, N. et al. (2005). *Nucl. Instr. and Meth.* **A537**, 537. Distribution of the total Čerenkov signal for 200 GeV multiparticle events (**c**) and the distributions for three subsets of events selected on the basis of the fractional contribution of neutrons to the scintillator signal (**d**). From: Akchurin, N. et al. (2009). *Nucl. Instr. and Meth.* **A598**, 422

A comparison between Fig. 8.11b and d shows that the signal distributions from the event samples are clearly wider when the total kinetic neutron energy is chosen to dissect the overall signal. This is consistent with our assessment that dual-readout is a more effective way to reduce the effects of fluctuations in invisible energy on the hadronic energy resolution.

Apart from that, dual-readout offers also several other crucial advantages:

- Its use is not limited to high-$Z$ absorber materials.
- The sampling fraction can be chosen as desired.
- The performance does not depend on detecting the neutrons produced in the absorption process. Therefore, there is no need to integrate the calorimeter signals over a large detector volume.
- The signal integration time can be limited for the same reason.

This is not to say that there is no advantage in detecting the neutrons produced in the shower development. In fact, this may further improve the hadronic calorimeter resolution, since $f_{em}$ and $f_n$ are correlated with the nuclear binding energy losses in different ways, and thus may offer complementary benefits. Figure 8.12a shows that a decrease in the Čerenkov/scintillation signal ratio (from which $f_{em}$ can be derived) corresponds to an increase of the neutron component ($f_n$) of the scintillation signal. However, as shown in Fig. 8.12b, this correlation is not perfect. In this scatter plot, the $f_n$ values are plotted for two narrow bins in the distribution of the

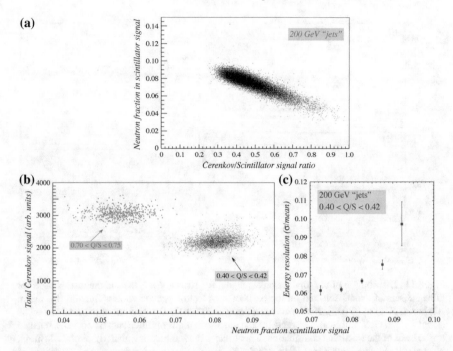

**Fig. 8.12** Results obtained with the DREAM copper-fiber dual-readout calorimeter, in which the time structure of the signals was measured [22]. A scatter plot in which the measured contribution of neutrons to the signals ($f_n$) is plotted versus the measured ratio ($Q/S$) of the Čerenkov and scintillation signals (**a**). A scatter plot in which the measured Čerenkov signal is plotted versus $f_n$, for two different bins of the $Q/S$ distribution (**b**). The energy resolution as a function of $f_n$, for events with (approximately) the same $f_{em}$ value (**c**). From: Akchurin, N. et al. (2009). *Nucl. Instr. and Meth.* **A598**, 422

Čerenkov/scintillation signal ratio. In both cases, the $f_n$ values cover a much larger range than the ±2% range of the $f_{em}$ values. Figure 8.12c shows that the energy resolution depends rather strongly on the chosen $f_n$ value, for a given value of $f_{em}$. RD52 has shown that the complementary information provided by measurements of $f_{em}$ and $f_n$ leads to a further improved hadronic energy resolution [22]. However, even without explicitly determining $f_n$, which involves measuring the time structure of each and every signal, the hadronic energy resolution that can be obtained with the dual-readout method is already superior to what has been achieved by the best compensating calorimeters.

## 8.5   Exploiting the Time Structure of the Signals

The availability of ultrafast electronics at a reasonable price has opened new possibilities for applications in calorimetry. In this subsection, we give some examples of how relatively modest time resolutions could help with particle identification and with the elimination of systematic sources of error in calorimetric measurements.

There is a deeply rooted belief that calorimeter systems for high-energy collider experiments should be longitudinally subdivided into several sections. As a minimum, one will usually want to have an electromagnetic and a hadronic section. A major reason for this belief is that such a subdivision is needed for recognizing em showers, and thus identify electrons and $\gamma$s entering the calorimeter.

This is a myth. It has been demonstrated repeatedly that there are several ways to identify em showers in longitudinally *unsegmented* calorimeters, and a good time resolution can be a wonderful tool in that respect. Figure 8.13 illustrates one of these methods, developed by the RD52 Collaboration, which uses the starting time of the calorimeter signals, measured with respect to the signal produced in an upstream

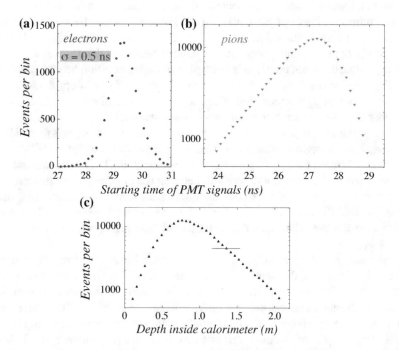

**Fig. 8.13** The measured distribution of the starting time of the calorimeter's scintillation signals produced by 60 GeV electrons (**a**) and 60 GeV pions (**b**). This time is measured with respect to the moment the beam particle traversed a trigger counter installed upstream of the calorimeter. These data were also used to determine the distribution of the average depth at which the light was produced in the hadron showers (**c**). The horizontal line represents the position resolution of this measurement. From: Akchurin, N. et al. (2014). *Nucl. Instr. and Meth.* **A735**, 120

detector [23]. This method is based on the fact that light in the optical fibers travels at a lower speed ($c/n$) than the particles that generate this light ($\sim c$). The deeper inside the calorimeter the light is produced, the earlier the calorimeter signal starts. For the polystyrene fibers used in this detector, the effect amounted to 2.55 ns/m.

Figure 8.13 shows the measured distribution of the starting time of the signals from 60 GeV $e^-$ (Fig. 8.13a) and $\pi^-$ (Fig. 8.13b). The starting time of the electron signals has a standard deviation of 0.5 ns. Since all em showers generated light at approximate the same depth inside the calorimeter, this time resolution translates into a longitudinal position resolution of $\sim$20 cm. This pion distribution peaked $\sim$1.5 ns earlier than that of the electrons, which means that the light was, on average, produced 60 cm deeper inside the calorimeter. The distribution is also asymmetric, it has an exponential tail towards early starting times, i.e., light production deep inside the calorimeter. This signal distribution was also used to reconstruct the average depth at which the light was produced for individual pion showers. The result, depicted in Fig. 8.13c, essentially shows the longitudinal profile of the 60 GeV pion showers in this calorimeter.

Apart from particle identification, the measurement of the depth of the light production in a longitudinally unsegmented calorimeter may also turn out to be useful for other purposes. For example, it may be used to correct for the effects of light attenuation in the fibers on the calorimeter signals. Even though the attenuation lengths are typically (considerably) longer than 5 m, uncertainties in the depth at which the light is produced are not completely negligible in high-resolution hadron calorimeters. The position resolution of 20 cm (indicated by the red line in Fig. 8.13c) limits the contribution of light attenuation effects to the energy resolution. It can of course be further improved when better time resolution is available.

It is also possible to use the time structure of the calorimeter signals themselves, for example for particle identification. This was demonstrated by the SPACAL Collaboration, who used the pulse width at 20% of the amplitude (FWFM) to this end [24] and measured very significant differences between the distributions of this variable for electrons and pions. In their case, the differences were considerably increased by the fact that the upstream ends of their fibers were made reflective. Therefore, the deeper inside the calorimeter the (scintillation) light was produced, the wider the pulse (Fig. 8.14).

As the available time resolution further increases, other applications may become feasible. For example, the depth measurement in several neighboring towers contributing to the shower signal may provide an indication of the *direction* at which the particle(s) entered the calorimeter, thus allowing measurement of the entire four-vector. And time resolutions of $\sim$10 ps might even make it possible to mitigate the effects of "pile-up", which could seriously deteriorate the performance of calorimeters in the high-luminosity era of the LHC operations [25].

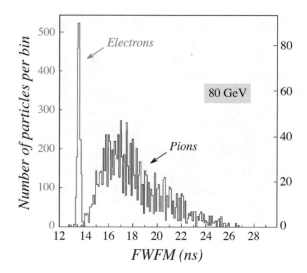

**Fig. 8.14** Distribution of the Full Width at one Fifth of the Maximum (FWFM) of the signals produced by 80 GeV electrons and pions in a longitudinally unsegmented fiber calorimeter. From: Acosta, D. et al. (1991). *Nucl. Instr. and Meth.* **A305**, 55

# References

1. Fabjan, C.W., et al.: Nucl. Instrum. Meth. **141**, 61 (1977)
2. d'Agostini, G., et al.: Nucl. Instrum. Methods **A274**, 134 (1989)
3. De Vincenzi, M., et al.: Nucl. Instrum. Methods **A243**, 348 (1986)
4. Bernardi, E., et al.: Nucl. Instrum. Methods **A262**, 229 (1987)
5. Galaktionov, Y., et al.: Nucl. Instrum. Methods **A251**, 258 (1986)
6. Wigmans, R.: Calorimetry—Energy Measurement in Particle Physics, 2nd edn. International Series of Monographs on Physics, vol. 168. Oxford University Press, Oxford (2017)
7. Akchurin, N., et al.: The response of CMS combined calorimeters to single hadrons, electrons and muons, CERN-CMS-NOTE-2007-012 (2007)
8. Acosta, D., et al.: Nucl. Instrum. Methods **A308**, 481 (1991)
9. Suzuki, T., et al.: Nucl. Instrum. Methods **A432**, 48 (1991)
10. Behrens, U., et al.: Nucl. Instrum. Methods **A289**, 115 (1990)
11. Lee, S., Livan, M., Wigmans, R.: Rev. Mod. Phys. **90**, 025002 (2018)
12. Abramowicz, H., et al.: Nucl. Instrum. Methods **180**, 429 (1981)
13. Akchurin, N., et al.: Nucl. Instrum. Methods **A537**, 537 (2005)
14. Particle Data Group, Tanabashi, M., et al.: Phys. Rev. **D98**, 030001 (2018); Section 34.9.2
15. Lee, S., et al.: Nucl. Instrum. Methods **A866**, 76 (2017)
16. Akchurin, N., et al.: Nucl. Instrum. Methods **A610**, 488 (2009)
17. Akchurin, N., et al.: Nucl. Instrum. Methods **A735**, 130 (2014)
18. Cardini, A., et al.: Nucl. Instrum. Methods **A808**, 41 (2016)
19. Lee, S., Livan, M., Wigmans, R.: Nucl. Instrum. Methods **A882**, 148 (2018)
20. Agostinelli, S., et al.: Nucl. Instrum. Methods **A506**, 250 (2003)
21. http://geant4.cern.ch/support/proc_mod_catalog/physics_lists/useCases.shtml
22. Akchurin, N., et al.: Nucl. Instrum. Methods **A598**, 422 (2009)
23. Akchurin, N., et al.: Nucl. Instrum. Methods **A735**, 120 (2014)
24. Acosta, D., et al.: Nucl. Instrum. Methods **A305**, 55 (1991)
25. Barnyakov, A.Yu., et al.: Nucl. Instrum. Methods **A879**, 6 (2018)

# Part IV
# Challenges Encountered in Practice

# Chapter 9
# Calibrating a Calorimeter System

## 9.1 Introduction

Calibration is an incredibly important aspect of working with calorimeters. Physicists want to measure particle energies with their calorimeters. The calorimeters produce electric signals. The calibration gives the recipe for converting one into the other.

While usually enormous efforts and ingenuity are invested in the design and the construction of calorimeters, their calibration is often a stepchild, left to be taken care of by someone who has nothing better to do. This rather general disregard for this aspect of calorimetry probably stems from the perception that the job is trivial and straightforward. In this chapter, we show that this is by no means the case.

In the previous chapters, several reasons why calibrating a calorimeter system is not a trivial and straightforward job have already been encountered:

1. The calorimeter response, defined as the *average signal per GeV*, depends on the *type* of particle, or the type of jet. The calibration constants, which are expressed in units of "pc/GeV" or "p.e./GeV" thus also depend on the type of particle, or jet. Calibration constants determined for one particular type of particle, or jet, thus lead to *systematic mismeasurements* of energy if used for the interpretation of signals caused by other types of particles, or jets. An example was shown in Fig. 5.12.
2. The calorimeter response, and therefore the calibration constants, may depend on the *energy* of the particles, or the jets. In that case, calibration constants determined for particles or jets of one particular energy lead to systematic mismeasurements of energy if used for particles or jets of other energies. An example was shown in Fig. 4.2.
3. The calorimeter response to showering particles is a function of the *shower age*. The em response of certain calorimeters may change by as much as 30% between the early and late stages of the shower development (Fig. 3.13). Even larger effects may occur in hadron showers.

© Springer Nature Switzerland AG 2019

M. Livan and R. Wigmans, *Calorimetry for Collider Physics, an Introduction*,
UNITEXT for Physics, https://doi.org/10.1007/978-3-030-23653-3_9

All these effects may thus lead to a *systematic mismeasurement* of the energy. This fact tends to be ignored. Calibration problems are most severe for longitudinally subdivided calorimeter systems, especially if the calorimeter consists of sections with very different $e/h$ values. However, before going into these problems, we first discuss the easier case of a longitudinally unsegmented calorimeter.

## 9.2  Longitudinally Unsegmented Systems

The best way to calibrate a longitudinally unsegmented calorimeter system is to send a beam of electrons with precisely known energy into the center of each and every calorimeter cell. If the cell size is such that the em showers are completely contained in one cell, the calibration constant of each cell is given by the ratio of the beam energy and the mean value of the signal distribution for that cell.

If the shower energy is only partially contained in one calorimeter cell, the calibration constants $C_i$ have to be determined simultaneously for all cells $i$, by minimizing the quantity

$$Q = \sum_{j=1}^{N} \left[ E - \sum_{i=1}^{n} C_i S_{ij} \right]^2 \tag{9.1}$$

in which $E$ denotes the beam energy and $S_{ij}$ the signal from calorimeter cell $i$ for event $j$.

In calorimeters where the signals are amplified, it is often convenient to choose the gain factors such that equal energy deposits lead to equal signals, in all calorimeter cells. In that way, signals from different cells may be directly added without applying normalization constants. This is particularly convenient when combinations of calorimeter signals from different cells are used to decide whether an event meets preset selection criteria (triggering).

If the calorimeter is compensating, the calibration constants derived from em shower detection are also valid for hadrons and jets. This is of course the most ideal situation.

If the calorimeter is non-compensating, energy-dependent correction factors need to be applied to derive the energy of hadrons and jets from the calorimeter signals generated by these objects. These correction factors can be established with test beams of hadrons and electrons, preferably spanning the entire energy range of interest in the experiment for which the calorimeter is intended.

## 9.3  Longitudinally Segmented Systems

Almost all calorimeter systems used in practice consist of several longitudinal segments. Most systems consist of two segments, named the *electromagnetic* and *hadronic* calorimeter sections, but three or even (many) more segments are not unusual either. Some calorimeters for PFA based systems[1] have more than 50 longitudinal segments [1].

Arguments typically used to design a calorimeter system in this way are:

1. It makes it possible to use available resources in the most economical way, by constructing a fine-sampling, high-granularity electromagnetic section, followed by a more crudely designed hadronic compartment.
2. It allows easy identification of electrons and $\gamma$s.
3. It helps to improve the hadronic performance.

Calibration complications are the price to pay for these (perceived) advantages. These complications are such that a proper alternative title of this chapter would be:

*(Mis)calibration – The pitfalls of longitudinal segmentation*

In some cases (segments with different $e/h$ values), it is even questionable if there is a correct way to calibrate longitudinally segmented calorimeters at all.

## 9.3.1 The Basic Problem

We start the description of these complications with a very simple case, involving em showers in a homogeneous calorimeter, which we assume to be based on the detection of Čerenkov light, e.g., a block of lead glass (Fig. 9.1). When electrons are sent into this detector, Čerenkov photons are generated in the absorption of the shower, and these photons are converted into photoelectrons in a light sensor, which produces the signals. When this device is calibrated with 100 GeV electrons, the signal from his sensor consists, on average, of 1000 photoelectrons. We conclude that the calibration constant for this calorimeter is 10 photoelectrons per GeV deposited energy (10 p.e./GeV, or 0.1 GeV/p.e.). Since this calorimeter is linear, a beam of 20 GeV electrons will produce a signal that, on average, consists of 200 photoelectrons ($20 \times 10$, or $20/0.1$). Since we have defined the calorimeter response as the average signal per GeV deposited energy, we can also say that the *response* of this calorimeter is 10 photoelectrons (per GeV), and the average signal for a 50 GeV electron will thus consist of $50 \times 10 = 500$ photoelectrons.

Now we are going to cut this detector into three parts, or rather we arrange things in such a way that the signals produced in segments I, II and III are detected separately (Fig. 9.1). This cut is made such that the 100 GeV electrons that were used for the detector calibration deposit, on average, 30% of their energy in segment I, 40% in segment II and 30% in segment III.

Čerenkov light is only produced by the charged shower particles that are sufficiently relativistic, e.g., the electrons and positrons that carry at least 0.3 MeV kinetic energy. Shower particles with energies below this cutoff value do participate in the energy deposition process, but *not* in the signal generation. These soft particles are

---

[1]PFA = *Particle Flow Analysis*, the topic of Chap. 12.

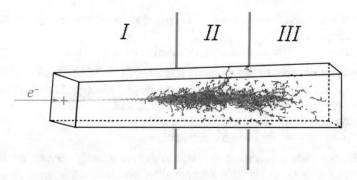

**Fig. 9.1** A hypothetical homogeneous lead glass calorimeter in which an electromagnetic shower develops. This detector is cut in three components, which are read out separately [2]

rather rare in the early stages of the shower development, but they dominate in the late stages. This means that if we now calibrate the three segments of the calorimeter separately, a different relationship will be found between deposited energy (in GeV) and resulting signal (in photoelectrons) for these three segments. For example, we find that in the first segment 15 photoelectrons are produced per GeV deposited energy. In segment II, the signals from the 100 GeV electron showers consist, on average, of 10 photoelectrons per GeV deposited energy, and in segment III the calibration constant is 5 photoelectrons per GeV.

With these new, separate calibration constants, the average total calorimeter signal for 100 GeV electron showers is still the same as before the cut was made. Because of the 30%/40%/30% sharing of the deposited energy, segment I will contribute on average $30 \times 15 = 450$ photoelectrons to the total signal, segment II $40 \times 10 = 400$ and segment III $30 \times 5 = 150$, for a total of $450 + 400 + 150 = 1000$ photoelectrons.

However, if we now send electrons of another energy into this segmented calorimeter, the energy sharing between these three segments will be different than for the 100 GeV ones, because of differences in the longitudinal shower profile (Fig. 3.1). For example, the energy sharing among segments I, II and III for a 20 GeV electron is, on average, 45%/35%/20%, i.e., 9 GeV in segment I, 7 GeV in segment II and 4 GeV in segment III. Based on the calibration constants of these segments established with the 100 GeV electrons, the energy of the 20 GeV electrons will be *underestimated*. This can be easily seen as follows. Imagine that the entire energy of the 20 GeV shower was deposited in segment I. The shower does not know that the calorimeter readout is split into three segments and would still produce 200 photoelectrons, as before. However, these would now be interpreted as an energy deposit of 200/15 = 13.3 GeV, because energy deposit in segment I is converted on the basis of 15 p.e./GeV. Given the mentioned energy sharing for 20 GeV showers, the energy would only be correctly reproduced if the shower produced $9 \times 15 + 7 \times 10 + 4 \times 5 = 225$ photoelectrons, instead of 200. The showers would thus be assigned an energy of $20 \times 200/225 = 17.8$ GeV in this example.

Similarly, the energy of electrons with an energy larger than 100 GeV would be systematically *overestimated* in this example, since they would deposit a relatively large fraction of their energy in segment III, which is more generous in converting photoelectrons into GeVs than the other segments. For example, a 500 GeV electron with energy sharing 20%/30%/50% = 100/150/250 GeV, would be reconstructed at the correct energy if it generated $100 \times 15 + 150 \times 10 + 250 \times 5 = 4{,}250$ photoelectrons. However, since the particles generate, on average, 5,000 photoelectrons, their signals would be interpreted as coming from a 588 GeV electron.

These problems are not limited to electrons with different energies than the one used for calibrating the individual longitudinal segments. Also a $\gamma$ of 100 GeV would be reconstructed, on average, with a different energy than a 100 GeV electron. This is a consequence of differences between the average longitudinal shower profiles (Sect. 3.1.1). Since $\gamma$ showers deposit their energy typically deeper inside the calorimeter than electron showers of the same energy, the energy of a 100 GeV $\gamma$ would thus be interpreted as coming from a higher-energy object, e.g., 105 GeV. Experiments that have developed elaborate calibration schemes for their longitudinally segmented em calorimeter, based on electron detection (e.g., ATLAS, Sect. 9.3.4) ought to keep this in mind when reconstructing $\gamma$ showers.

This example illustrates the problems caused by item #3 listed in the introduction of this chapter, namely the fact that the calorimeter response (i.e., the relation between deposited energy and resulting signal) changes as the shower develops, and is thus a function of the depth inside the calorimeter. This implies that the calibration constants, which relate signals to deposited energy, are different for the different segments of a longitudinally segmented calorimeter. And since the shower profiles change with the particle energy, this leads to signal non-linearities. This is a very serious problem, as we will see in the following sections, which deal with real, non-hypothetical calorimeters.

## 9.3.2 The HELIOS Calorimeter

The calorimeter system for the HELIOS experiments in the heavy-ion beam at CERN [3] consisted of thin uranium plates interleaved with 3 mm thick plastic scintillator sheets. It was longitudinally subdivided into two sections. The first (em) section was 6.4 radiation lengths deep and used 2 mm thick uranium plates. The second (hadronic) section was ~4 nuclear interaction lengths deep. The uranium plates were 3 mm thick in that section. In the transverse plane, the calorimeter consisted of rectangular towers, measuring $20 \times 20$ cm$^2$, read out from two sides by means of wavelength shifting plates.

This calorimeter system was calibrated with electrons of 8, 17, 24, 32 and 45 GeV, and also with cosmic muons. When the electron beam was steered into the center of a calorimeter tower, showers were completely contained in one of the towers, with comparable fractions of the energy deposited in the two longitudinal segments. This is graphically shown in Fig. 9.2.

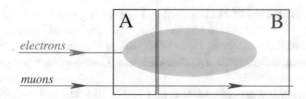

**Fig. 9.2** The HELIOS modules were calibrated with beams of electrons [3]. The showers produced by these particles produced signals of comparable strength in both sections of the calorimeter modules. Alternatively, the modules were calibrated with muons, which traversed the entire module [2]

The calibration constants $A$ and $B$ for the em and hadronic calorimeter sections, respectively, were determined by minimizing the width of the total signal distribution, i.e., by minimizing the quantity

$$Q = \sum_{j=1}^{N} \left[ E - A \sum_{i=1}^{n} S_{ij}^{\text{em}} - B \sum_{i=1}^{n} S_{ij}^{\text{had}} \right]^2 \qquad (9.2)$$

where $E$ is the electron beam energy and $\sum S^{\text{em}}$ and $\sum S^{\text{had}}$ are the sums of all the signals in the towers $i$ of the em and hadronic calorimeter sections that contributed to the measured signal for event $j$. With this method, values for $A$ and $B$ and, more importantly, for the intercalibration constant $B/A$ were determined for each calorimeter tower.

However, two fundamental difficulties were encountered when this calibration method was applied:

1. The values of $A$, $B$ and $B/A$ were found to be energy dependent.
2. The values of $B/A$ differed considerably (on average, more than 20%) from the ones found with muons.

This is illustrated in Fig. 9.3, where the fractional widths of various total signal distributions, $\sigma_{\text{rms}}/E$, are plotted as a function of the value of $B/A$ (the back/front weighting factor). In the following, we use $B/A$ values that are normalized to the one for muons that traversed both calorimeter sections. The value $B/A = 1$ (indicated by the red dashed line in the figure) thus represents the calibration result derived from the muon signals, as described below. If $B/A > 1$, then the signals from the hadronic section were given a relatively larger weight for the calculation of the total energy. If $B/A < 1$, then the signals from the em section were given a larger weight.

The $B/A$ value for muons, and thus the normalization factor for all measurements, was experimentally determined as follows by the HELIOS group. Cosmic muons traversing both sections of a given tower were selected by means of a cosmic-ray telescope. These muons generated the characteristic Landau signal distributions in the em and hadronic sections of the tower. The $B/A$ value was determined from the ratio of the most probable signal values in the hadronic and em sections of the

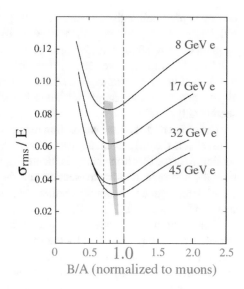

**Fig. 9.3** The fractional width $\sigma/E$ of the signal distributions for electrons of different energies, as a function of the value of the intercalibration constant $B/A$ of the HELIOS calorimeter system. The red dashed line corresponds to the intercalibration constant derived from muon measurements. From: Åkesson, T. et al. (1987). *Nucl. Instr. and Meth.* **A262**, 243

tower. This ratio was compared with its expected value. The latter did not require an experimental measurement, but could be calculated from the composition of the calorimeter and from the specific ionization of the active and passive materials for mips traversing both sections. The overall normalization factor was chosen such as to equalize the measured and expected signal ratios. This procedure was repeated for each individual calorimeter tower.

Figure 9.3 shows that the fractional widths of the signal distributions for electrons reached a minimum for $B/A$ values well below the value expected on the basis of the muon measurements ($B/A = 1.0$). In addition, the $B/A$ value for which this width reached a minimum shifted upward with the electron energy, by about 20% over the energy range for which measurements were done (8–45 GeV).

The HELIOS group also found that using $B/A$ values different from 1.0 resulted in signal non-linearity: the em response became energy dependent. Signal linearity for em showers is a very fundamental calorimeter property, since the entire em shower energy is used to ionize and excite the molecules of which the calorimeter consists. Twice as much energy thus leads to twice as many excited and ionized molecules and should thus lead to calorimeter signals that are twice as large. In the HELIOS case, it seemed impossible to fulfill this fundamental requirement and to optimize the energy resolution for em showers simultaneously.

The explanation for the described phenomena lies in the fact that the sampling fraction, i.e., the fraction of the deposited shower energy that is converted into a measurable signal, changes with depth (Sect. 3.5.2). In the (HELIOS) case of uranium absorber and plastic scintillator as active material, the sampling fraction *decreases* considerably as the shower develops, by as much as 25–30% over the volume in which the absorption takes place. In the early stage of its development, the shower still resembles a collection of mips, but especially beyond the shower maximum the energy is predominantly deposited by soft ($<1$ MeV) $\gamma$s. The latter are much less efficiently sampled than mips in this type of structure, where dominant processes such as photoelectric effect and Compton scattering strongly favor the high-$Z$ absorber material. Therefore, a given energy deposit by the fast (more mip-like) component of the shower in the electromagnetic calorimeter section leads to a considerably larger signal than the same energy deposited in the soft tail in the hadronic calorimeter section (Fig. 3.13).

The effect of this on the calibration results is energy dependent. The optimal $B/A$ ratio reflects the difference in (average) sampling fractions between both calorimeter sections. As the electron energy is increased, more energetic shower particles penetrate the hadronic calorimeter section and the response *difference* between the two calorimeter segments becomes smaller, i.e., the optimal value of the $B/A$ weighting factor gets closer to 1 (Fig. 9.3).

This phenomenon indicates that there is a fundamental problem. The relationship between the energy deposited by the shower particles and the resulting calorimeter signal (i.e., the calorimeter response) is very different for the two sections of the calorimeter and, moreover, energy dependent. Translating the signals in the individual calorimeter sections into deposited energy is therefore an extremely delicate issue, no matter which calibration constants are being used.

Based on the described phenomena, HELIOS decided to use the $B/A$ values derived from the muon measurements as the basis of their calorimeter calibration. The values of $A$ (and thus automatically of $B$) were fixed such as to reproduce best the electron beam energies over the measured range. In this way, signal linearity was achieved.

In the HELIOS calorimeter, the thickness of the electromagnetic calorimeter section was chosen such as to achieve roughly equal energy sharing for em showers. It turned out that this choice *maximized* the described effects. The hadronic energy resolution was much less sensitive to the value of $B/A$, because typically only a very small fraction of the shower energy was deposited in the front section.

### 9.3.3   Intercalibration with em Showers

There are some very important lessons to be learned from this HELIOS experience:

- Showers should *not* be used to intercalibrate sections of a longitudinally segmented calorimeter.
- In a longitudinally segmented calorimeter, no matter how the segments are intercalibrated, only the *total* shower energy derived from the signals is meaningful. The signals from the individual calorimeter segments cannot be interpreted in a straightforward way in terms of deposited energy.

Reference [2] describes a detailed Monte Carlo study of the effects of a calibration procedure in which the width of the total signal distribution of showers that develop in several different calorimeter segments is minimized. It is shown that such a procedure leads *inevitably* to a non-linear response. Results are given for various values of the intercalibration constant (or back/front weighting factor) $B/A$ for a calorimeter with the structure of the HELIOS one. Defining the degree of non-linearity as the fractional change of the electron response in the energy interval from 1 to 100 GeV, it was found to range from $-15\%$ for $B/A = 0.7$ to $+5\%$ for $B/A = 1.1$. Only for $B/A = 1$, was the detector found to be linear for electron detection.

Now one might argue that there is in principle no reason why a calorimeter that is non-linear for em shower detection, although somewhat inconvenient, should be unacceptable. Any type of non-linearity could in principle be dealt with by means of a polynomial relationship between the signals $S$ and the corresponding energy $E$:

$$E = c_0 + c_1 S + c_2 S^2 + c_3 S^3 + \ldots \tag{9.3}$$

and the fact that other constants than $c_1$ have a non-zero value might be a small price to pay for improving energy resolution. This line of reasoning is, however, crucially flawed (see also Sect. 4.2).

This is because the average energy sharing between the calorimeter sections (and thus the response) does not only depend on the particle's energy but also on the particle(s) that caused the shower. For example, the sharing is different for electrons and $\gamma$s of the same energy (Sect. 3.1.1). A $\gamma$ will deposit, on average, a larger fraction in the deeper calorimeter section than an electron of the same energy. If the signals from that section are given a smaller weight (which is what is happening for a calibration scheme with $B/A < 1$), then the response for such $\gamma$s will be smaller than for electrons. Also, if the em shower consists of several unresolved showers caused by lower-energy particles, then the average energy fraction in the deeper calorimeter section will be smaller, and this leads to an increased response. This may occur, for example, when a $K^0$ decays into several $\pi^0$s, each of which in turn decays into 2 $\gamma$s.

This is illustrated in Fig. 9.4, which shows these effects for three chosen calibrations ($B/A = 0.7$, 0.8 and 0.9, respectively). The mean response values differed by as much as 1 GeV. Only for $B/A = 1$ all distributions had, within the statistical uncertainty of the Monte Carlo simulations ($\sim 0.05$ GeV), the same mean value [4].

These effects are not small, given the fact that some of the calorimeters in particle physics experiments are nowadays designed to achieve sub-1% energy resolution. Figure 4.2 shows the signal distributions for 50 GeV $\gamma$s, 50 GeV $\pi^0$s and 50 GeV $K^0$s (decaying into $\pi^0\pi^0\pi^0$) in the simulated calorimeter, calibrated on the basis of $B/A = 0.8$ with 50 GeV electrons, and assuming an intrinsic energy resolution of 0.5% (i.e., $3.5\%/\sqrt{E}$). Based on this calibration, single photons were reconstructed with an energy that was, on average, too low by 0.67 GeV, i.e., 2.7 times the intrinsic energy resolution of the calorimeter. On the other hand, the energy of the kaons was systematically *overestimated*, on average by 0.85 GeV (3.4 $\sigma$). The reconstructed energy of the $\pi^0$s was approximately correct, because the energy sharing between the two calorimeter segments was approximately the same for showers of 25 GeV $\gamma$s (the decay products of the $\pi^0$s) and 50 GeV electrons (used for the calibration).

This analysis illustrates a fundamental problem inherent to non-linear calorimeters. The calorimeter information is intended to determine the energy of particles

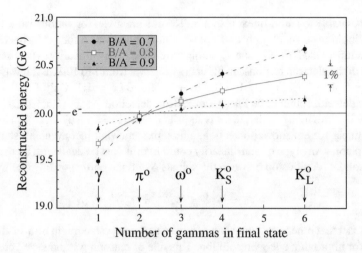

**Fig. 9.4** The average reconstructed energy for 20 GeV $\gamma$s and for 20 GeV particles decaying into multiple $\gamma$s in a longitudinally segmented Pb/scintillator calorimeter that was calibrated with 20 GeV electrons, for different values of the ratio of the calibration constants for the two longitudinal segments, $B/A$. See text for details. From: Wigmans, R. and Zeyrek, M. (2002). *Nucl. Instr. and Meth.* **A485**, 385

from the signals they generate. The precision with which the energy can be measured is determined by the energy resolution. However, in a non-linear calorimeter, *the energy resolution needed in this context does not correspond to the width of the signal distribution measured for single, monoenergetic particles*. This observation is also particularly important for the evaluation of jet resolutions in a non-compensating calorimeter.

### 9.3.4   Three Compartments—The ATLAS LAr Calorimeter

The calibration problems described in the previous subsections become more complicated when the calorimeter consists of more than two longitudinal segments. A recent example of an experiment that has to deal with this intercalibration issue is ATLAS, whose Pb/LAr electromagnetic calorimeter consists of three longitudinal segments. At $\eta = 0$, the depths of these segments are $4.3X_0$, $16X_0$ and $2X_0$, respectively. When the particles enter the barrel calorimeter at a non-perpendicular angle, the total depth of this calorimeter increases (from $22X_0$ at $\eta = 0$ to $30X_0$ at $|\eta| = 0.8$), and so do the depths of these three segments. The sampling fraction for mips is the same in all three segments.

Figure 9.5 shows how the sampling fraction for em showers evolves as a function of depth, in an energy dependent way. The sampling fraction for muons that traverse this detector does not change in this process and, therefore, the $e/mip$ value decreases

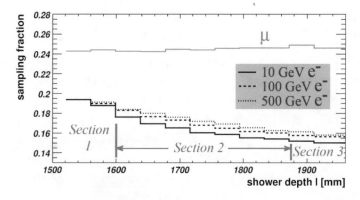

**Fig. 9.5** The evolution of the sampling fraction for electron showers of different energies in the three longitudinal segments of the ATLAS LAr calorimeter, at $\eta = 0$. From: Aharouche, M. et al. (2006). *Nucl. Instr. and Meth.* **A568**, 601

by 20–25%, depending on the electron energy. In light of our discussions in the earlier sections of this chapter, these data should by now look very familiar. Not surprisingly, the problems encountered when calibrating this detector with electron showers were also very similar to the ones experienced by HELIOS, except that there were now three calibration constants to be determined, instead of two. When minimizing $Q$ in

$$Q = \sum_{j=1}^{N} \left[ E - A \sum_{i=1}^{n} S_1^{ij} - B \sum_{i=1}^{n} S_2^{ij} - C \sum_{i=1}^{n} S_3^{ij} \right]^2 \tag{9.4}$$

it turned out that the resulting calibration constants $A$, $B$ and $C$ not only depended on the electron energy, as in HELIOS, but also on the location of the calorimeter module (the $\eta$ value). The latter dependence can be understood from the fact that the effective depth of the longitudinal segments changes with the angle of incidence of the particles. And just as in HELIOS, it was found that any choice of the calibration constants resulting from such a minimization procedure introduced a response non-linearity.

Rather than intercalibrating the different longitudinal calorimeter segments with muons (the HELIOS solution), ATLAS decided to approach this problem in a much more complicated way, relying heavily on Monte Carlo simulations in an attempt to achieve the best possible combination of energy resolution and linearity. These elaborate simulations led to a very complicated procedure for determining the energy of a shower detected in the various segments of the calorimeter. This procedure was based on a variety of parameters that depended both on the energy and the $\eta$ value (Fig. 9.6). It was tested in great detail with Monte Carlo events and combines excellent signal linearity with good energy resolution [5].

The energy dependence of the various parameters derives from the fact that the longitudinal shower profiles, and thus the energy sharing between the three segments,

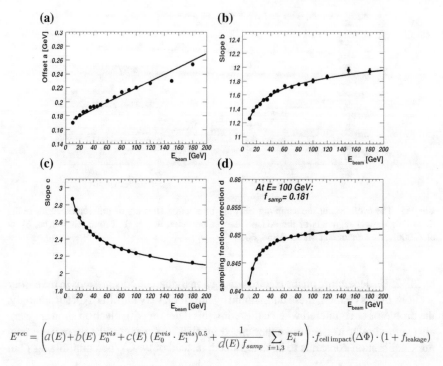

$$E^{rec} = \left(a(E) + b(E)\, E_0^{vis} + c(E)\,(E_0^{vis} \cdot E_1^{vis})^{0.5} + \frac{1}{d(E)\, f_{samp}} \sum_{i=1,3} E_i^{vis}\right) \cdot f_{\text{cell impact}}(\Delta\Phi) \cdot (1 + f_{\text{leakage}})$$

**Fig. 9.6** The formula used by ATLAS to determine the energy of a shower developing in the longitudinally segmented ECAL. The energy dependence of the various parameters is shown in graphs (**a–d**). From: Aharouche, M. et al. (2006). *Nucl. Instr. and Meth.* **A568**, 601

change with the energy of the incoming electron. In Sect. 3.1.1, we argued that showers induced by photons are in this respect quite different from showers induced by electrons (see for example Fig. 3.3). Figure 9.4 shows the consequences of that for the energy measurement. Therefore, we believe that the calibration procedure developed by ATLAS on the basis of electron showers (Fig. 9.6) would need to be modified for the detection of $\gamma$s and particles decaying into several $\gamma$s. In a recent paper [6], the ATLAS Collaboration describes a LAr calibration scheme based on experimental data and simulations that takes this issue into account.

### 9.3.5   Many Compartments—The AMS Calorimeter

The last example of the pitfalls of calibrating a longitudinally segmented device concerns the calorimeter for the AMS-02 experiment, which detects high-energy electrons, positrons and $\gamma$s at the International Space Station [7]. This calorimeter has eighteen independent longitudinal depth segments. Each layer consists of a lead absorber structure in which large numbers of plastic scintillating fibers are embedded, and is about $1X_0$ thick. A minimum ionizing particle deposits, on average, 11.7 MeV upon traversing such a layer. The AMS-02 collaboration initially calibrated this calorimeter by sending a beam of muons though it and equalizing the signals from all eighteen longitudinal segments. This seems like a very good method to calibrate this detector, since all eighteen layers have exactly the same structure. However, when this calorimeter module was exposed to beams of high-energy electrons, it turned out to be highly non-trivial how to reconstruct the energy of these electrons. Figure 9.7a shows the average signals from 20 GeV electron showers developing in this calorimeter. These signals were translated into energy deposits based on the described calibration. The measured data were then fitted to a $\Gamma$-function:

$$dE/dt \propto t^{\alpha} \exp(-\beta t)$$

where $t$ is the layer number and $\alpha$ and $\beta$ the coefficients to be fitted. Since the showers were not fully contained, the average leakage was estimated by extrapolating this fit to infinity. As shown in Fig. 9.7b, this procedure systematically underestimated this leakage fraction, more so as the energy (and thus the leakage) increased. The reason for this is the same as the one that plagued the calorimeters discussed in the previous subsections, namely the fact that the sampling fraction of em showers decreases as the shower develops. Therefore, a procedure in which the relationship between measured signals and the corresponding deposited energy is assumed to be

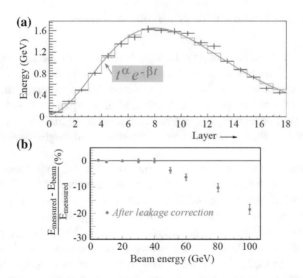

**Fig. 9.7** Average signals for 20 GeV electrons in the eighteen longitudinal sections of the AMS-02 lead/scintillating fiber calorimeter. The curve represents the results of a fit to the experimental data. The superimposed histogram is the expected average profile from the Monte Carlo simulations. Average relative difference between the measured energy and the electron beam energy, after leakage corrections based on extrapolation of the fitted shower profile were applied. From: Cervelli, F. et al. (2002). *Nucl. Instr. and Meth.* **A490**, 132

the same for each depth segment will cause the energy leakage to be systematically underestimated, more so if that leakage increases.

Based on these observations, AMS had to change its calibration procedure. Rather than just integrating the $\Gamma$-function that describes the measured shower profile to infinity, the energy of the particle that caused the shower is now determined on the basis of the fraction of the total signal measured in the last two detector segments. The deposited energy in the calorimeter derived from the $\Gamma$-function is subsequently corrected with a factor that is determined by this fraction. The value of this correction factor (needed to reproduce the actual electron energy) was established empirically in testbeam measurements [8], for the available energy range (up to 250 GeV).

The very complicated issues discussed here will most definitely also affect PFA calorimeters, which are all based on structures that are highly segmented, both longitudinally and laterally. The underlying problem is that the relationship between deposited energy and resulting signal is not constant throughout a developing shower. As the composition of the shower particles changes, so does the sampling fraction. The examples of the calorimeters discussed here clearly illustrate the problems that may be expected when this is not properly recognized and dealt with. The assumption that the relationship between deposited energy and recorded signal remains the same throughout the developing shower has been the modus operandi for the calorimeters built in the context of the CALICE project [1].

### 9.3.6   Intercalibration with Hadronic Showers

In the previous subsections we saw that intercalibration of the different compartments of a longitudinally segmented calorimeter leads to fundamental problems, even for em showers in compensating calorimeters. These problems only get worse when the intercalibration is carried out with *hadron* showers, in *non-compensating* calorimeters. Yet this technique is in practice frequently used to determine the calibration constants of the em and hadronic compartments of longitudinally segmented calorimeters.

One experiment that used this technique was the H1 experiment at HERA [9]. A prototype of their liquid-argon calorimeter consisted of a 1.06 $\lambda_{int}$ deep em section with lead absorber, followed by a hadronic section based on copper absorber ($6.12\lambda_{int}$) and a tail catcher with iron absorber ($2.88\lambda_{int}$). It turned out that the calibration constants obtained with a procedure minimizing the total width of the signal distribution did not only depend on the energy, but also on the type of particles used for this purpose. In the final instrument, they further increased the number of longitudinal segments and used an alternative method known as "offline compensation" (Sect. 9.5).

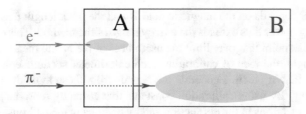

**Fig. 9.8** Calibration of the different longitudinal compartments of a calorimeter system with particles that deposit (almost) their entire energy in one of the compartments. Electrons are used to calibrate the em section, while pions penetrating this section without starting a shower are used to calibrate the hadronic calorimeter section [2]

### 9.3.7 Each Section Calibrated with Its Own Particles

As illustrated by the discussion in the previous subsections, intercalibration of the longitudinal segments of a calorimeter by means of showering particles that deposit part of their energy in each of the different segments is not a good idea. So what are the alternatives?

Many calorimeter systems consist of two sections, an electromagnetic section that is deep enough to contain showers induced by electrons and photons at the 95+% level (typically $\sim 25 X_0$), and a hadronic section in which hadrons typically deposit most of their energy. One method that is frequently used to calibrate such calorimeters is described in this subsection, and graphically depicted in Fig. 9.8.

In this method, both longitudinal calorimeter segments are *individually* calibrated. The calibration constants for the calorimeter cells of the em section are determined with beams of monoenergetic electrons. These electrons deposit their energy entirely in this section and, therefore, the same method as described in Sect. 9.2 for longitudinally unsegmented devices can be applied to determine the calibration constants for the individual calorimeter cells that make up the em calorimeter compartment.

The cells of the hadronic calorimeter section are calibrated with beams of hadrons, but only those events in which the beam particles penetrate the em calorimeter section without undergoing a strong interaction are selected for this purpose [10]. Apart from a very small fraction of energy lost through electromagnetic interactions in the front section (typically a few hundred MeV), the hadrons dump their entire energy in the hadronic calorimeter section. Calibration constants for the individual hadronic cells are determined in the same way as described in Sect. 9.2 for longitudinally unsegmented calorimeters.

This method may also be used in situ, for example in collider experiments with a central magnetic field (e.g., ATLAS and CMS at the LHC). Tracks from isolated charged particles that penetrate the em calorimeter section may be selected for that purpose. The momentum of the particles is given by the curvature of the tracks in the magnetic field. For large momenta, the deposited energy is almost independent of the type of particle. This type of calibration works of course better when the magnetic fields are strong and extend over a large volume, since the accuracy of the momentum measurement depends on the magnetic field $B$ and the track length $\ell$ as $(B\ell^2)^{-1}$. A large bending power thus extends the energy range of the method. This energy range is often a weak point that may limit the usefulness of the in situ measurements.

At first sight, the idea of calibrating each calorimeter section separately, using the particles that in practice generate the signals from these individual sections, is very appealing. By using calibration constants that correctly reproduce the energy deposited by electrons in the em section and by hadrons in the hadronic section, one hopes to eliminate the problems arising from the fact that the calorimeter response to these particles is different in non-compensating calorimeters. In all calorimeters with $e/h > 1$, the response to hadrons is smaller than that to electrons of the same energy. For such calorimeters, the described method thus corresponds to a choice of $B/A > 1$. In other words, if electrons were to be sent into the hadronic section of a calorimeter calibrated this way, their energy would be overestimated, on average by factor of $B/A$.

This calibration method works fine for those hadrons that penetrate the em section without starting a shower. However, this sample usually represents only a small fraction of all hadrons. Most hadrons undergo their first nuclear interaction in the em calorimeter section. They deposit (a sometimes large) part of their energy in the em section and the remainder in the hadronic section. For these events, this calibration method does not produce correct results.

It turns out that in non-compensating calorimeters (with $e/h > 1$), the energy of early showering particles is systematically underestimated as a direct result of this calibration method, to an extent that depends on the energy sharing between the two calorimeter compartments. This calibration method thus introduces a strong dependence on the starting point of the showers. Hadrons that happen to interact early in the detector are attributed an energy that is, on average, considerably lower than for hadrons that start showering after penetrating an interaction length or more.

A second undesirable consequence of this calibration method concerns the signal linearity for hadrons. As discussed in Sect. 7.2, all non-compensating calorimeters are intrinsically non-linear for hadron detection, since the average energy fraction carried by the em shower component depends on the energy of the incoming particle (Fig. 7.2a). However, this calibration method tends to make the non-linearity worse. An example of this effect is shown in Fig. 9.9b.

This can be understood as follows. As the energy of the incident hadron increases, a larger fraction of the shower energy is deposited in the hadronic calorimeter section. Since the signals from this section are amplified (by a factor $B/A$, see above), the hadronic response increases with energy as a direct result of this calibration procedure. The intrinsic non-linearity, i.e., the increase of the hadronic response because of the increasing fraction of the shower energy carried by $\pi^0$s, is thus further enhanced.

Although the calibration method discussed in this subsection looked at first sight attractive and logical, it thus turned out to be fundamentally flawed. It should be emphasized that the effects discussed here are an artifact of all calorimeters with $e/h \neq 1$. The degree to which the reconstructed hadron energy depends on the starting point of the shower development and the degree to which the non-linearity of the hadronic response are deteriorated as a direct consequence of applying this calibration method depend on the $e/h$ values of both the em and the hadronic calorimeter sections and on the thickness of the em section, expressed in $\lambda_{\text{int}}$.

Only when $e/h = 1.0$ for both calorimeter sections are the effects described above avoided, because in that case this calibration method is equivalent to using $B/A = 1$.

## 9.3.8 Forcing Signal Linearity for Hadron Detection

The third calibration method that we will discuss was inspired by the desire to eliminate the non-linearities in the hadronic response that are characteristic for non-compensating calorimeters. In this method, which is described and applied in [11], the signals from the em section are given different weights for the detection of electrons and hadrons. The calibration constants for em showers are determined in the same way as described in the previous subsection, i.e., by means of electrons of known energy that deposit all their energy in the em calorimeter section. However, when hadrons are showering in the calorimeter, the signals from the em section are weighted with a factor that is chosen to achieve signal linearity for these particles.

Usually, non-compensating calorimeters have an $e/h$ value larger than 1. In such *undercompensating* calorimeters, the hadronic response increases with energy (Fig. 7.2a). As the energy increases, the fraction of the shower energy deposited in the hadronic calorimeter section increases, on average, as well. Therefore, restoring linearity for hadrons in such calorimeters may be achieved either by suppressing the signals from the hadronic compartment or by boosting the signals from the em compartment. This is equivalent to applying a weighting factor to the intercalibration constant $B/A$ that is smaller than 1.

When hadrons are detected in an under-compensating calorimeter that is calibrated in this way, the signals from the em compartment will thus be attributed a larger energy than for electrons that produce the same signals in the same compartment.

As a result, the hadronic signal distributions obtained on the basis of this calibration method turn out to depend even more on the starting point of the showers than those based on the calibration scheme from the previous subsection. In particular, they depend on the fraction of the shower energy deposited in the em section. In extreme cases, differences of a factor of three in the response have been reported for hadrons of the same energy [12]. Especially for jets with a leading $\pi^0$, a large fraction of the energy is deposited in the em calorimeter section. As a result, the energy of such jets is typically *overestimated by a substantial factor* when this calibration method, defined on the basis of the response to single hadrons, is used.

This calibration method was developed to eliminate the energy dependence of the hadronic calorimeter response. It succeeded in achieving that goal. However, there is a price to be paid, as a direct consequence of this method:

- The hadronic response becomes dependent on the starting point of the showers.
- The response becomes dependent on the longitudinal shower development characteristics. This leads to differences between the calorimeter responses to jets and to single hadrons.

As before, these effects are avoided when $e/h = 1.0$ for both calorimeter sections, because in that case this calibration method would be equivalent to using $B/A = 1$.

### 9.3.9  No Starting Point Dependence of Hadronic Response

The fourth calibration method is, just like the three previous ones, inspired by a laudable goal. The purpose of this method is to make the hadronic response independent of the starting point of the shower. It was used by CDF for the calibration of their "plug upgrade" calorimeter [13]. CDF also tried two other calibration methods and compared the results, some of which are shown in Fig. 9.9. The method that is the topic of this subsection is called "Method III". Method I was based on the calibration procedure described in Sect. 9.3.7, where the calibration constant ("$B_I$") of the hadronic compartment[2] was determined by the response to pions that penetrated the em compartment without starting a shower. As shown in Fig. 9.9a, this calibration constant increased with the pion energy, because of the increased average em shower component. This was not the case with Method II, in which the calibration constant of the hadronic compartment was determined in the same way as for the em compartment, namely with electrons. And since both compartments were linear for em shower detection, both $A_{II}$ and $B_{II}$ were independent of the particle energy. This is in essence the method discussed in Sect. 9.3.11, and which we call the "B/A = 1" method. CDF only used Method III to determine the calibration constants with pions of 10 GeV, because of concerns that shower leakage might affect the results at higher energies.

Figure 9.9b shows the response to pions as a function of energy, using *all* events, i.e., not only the pions that deposited their entire shower energy in the hadronic compartment. For all three methods, the pion response increased with energy, as a result of the fact that $e/h > 1$ in both calorimeter compartments. However, the non-linearity was clearly worst for Method I. As the pion energy increased, a larger fraction of the shower energy was deposited in the hadronic calorimeter compartment. Boosting the signals from that compartment, which is the essence of Method I, thus tended to increase the already existing non-linearity. For Method III, the opposite effect occurred. By giving less weight to the signals from the hadronic compartment, the "natural" response non-linearity was reduced, albeit it not to the extent seen in Sect. 9.3.8, where the suppression of the signals from the hadronic compartment was so large that it resulted in response linearity.

---

[2]Note that CDF expressed calibration constants in ADC cts/GeV, whereas we use the inverse quantity (GeV/ct) elsewhere in this chapter.

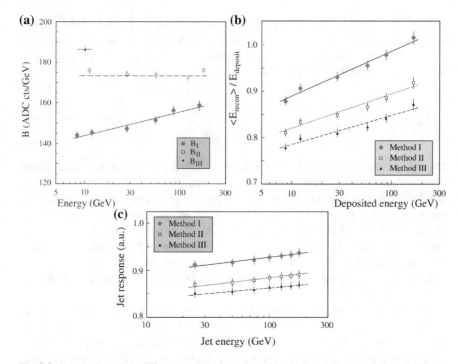

**Fig. 9.9** Results from three different methods investigated by CDF to calibrate their forward ("plug upgrade") calorimeter. The methods are described in the text. Shown are the calibration constants for the hadronic calorimeter compartment (**a**) and the response to single pions (**b**) and to jets (*c*) as a function of energy. From: Albrow, M. et al. (2002). *Nucl. Instr. and Meth.* **A487**, 381

CDF also studied the effects of these three calibration methods for jets, using a semi-empirical procedure to determine the jet response. Figure 9.9c shows the calorimeter response to these jets, as a function of energy. The same trends are observed as for single pions, but the non-linearities are clearly smaller for all methods. This is due to two factors:

1. Part of the jet signal, on average one-third, comes from $\gamma$s which develop em showers. The calorimeter is linear for that component of the jet signal. The non-linearity only affects the remaining portion of the jet fragments.
2. The non-linearity is not determined by the jet energy itself, but by *the average energy of the jet fragments*. Since the multiplicity increases with energy, this average energy of the fragments increases more slowly than the jet energy itself.

Yet, the differences between the non-linearities observed for the three calibration methods indicate that the effects described for single pions propagate into the energy measurement of jets.

Even though Method III eliminates the starting point dependence of the hadronic response, it introduces at the same time other problems. For example, the response to muons becomes different for the two compartments of the calorimeter system, to an extent determined by the difference between the $e/h$ values of these compartments. In today's experiments, the em calorimeter typically has a larger $e/h$ value than the hadronic compartment, since this is a general consequence of a larger sampling fraction (see, for example, Fig. 8.2).

### 9.3.10  Dummy Compensation

The calibration methods discussed in Sects. 9.3.7–9.3.9 tried to deal, in different ways, with the consequences of non-compensation ($e/h \neq 1.0$) for the energy scale of calorimeter systems. We briefly want to mention one other method that has been proposed for that purpose [14, 15]. The authors proposed to *eliminate* the effects of non-compensation by installing an inactive absorber upstream of the calorimeter. This "dummy section" would absorb, on average, a larger fraction of the energy of incident electrons than for hadrons. For an appropriate thickness of this absorber ($\sim 8X_0$), the average calorimeter signals for hadrons and electrons would then be equal, at least for one energy.

Such an arrangement would correspond to a calibration with $B/A = \infty$, since no signals from the first (dummy) section would be recorded. The main problem introduced by this method is a general deterioration of the calorimeter performance. Only signals corresponding to the shower energy deposited beyond a certain depth in the absorber structure would be recorded in this scheme. By ignoring the fluctuations in the energy deposited in the first $8X_0$ of the absorber structure, the precision of the information that could be obtained on the particles that initiated the showers inevitably deteriorates.

With regard to calibration, this method suffers from the same problems as the other calibration schemes based on $B/A \neq 1$:

- The hadronic response becomes dependent on the starting point of the showers.
- The hadronic response is different for jets and single hadrons with the same em shower content.
- The hadronic signal non-linearity is increased for calorimeters with $e/h > 1$.

In addition, the dummy section makes the calorimeter non-linear for electromagnetic showers. This calibration method thus combines the negative aspects of *all* other methods discussed in the previous subsections.

### 9.3.11  The Right Way

In the previous subsections, we have encountered a number of different methods that are being used to intercalibrate the signals from the different compartments of a longitudinally segmented calorimeter systems. All these methods were based on a specific goal that, at least at first sight, looked quite reasonable and is briefly summarized below:

1. Minimization of the width of the total signal distribution (Sect. 9.3.6)
2. Correct energy reconstruction of pions penetrating the em compartment without starting a shower (Sect. 9.3.7)
3. Hadronic signal linearity (Sect. 9.3.8)
4. Independence of hadron response on starting point shower (Sect. 9.3.9)
5. Equal response to electrons and pions (Sect. 9.3.10)

We also saw that each of these approaches introduced specific additional problems. We now conclude this section on the calibration of longitudinally segmented calorimeter systems with a statement that could summarize this entire chapter and therefore deserves to be printed in bold face.

**The correct way to intercalibrate the different sections of a longitudinally segmented, non-compensating calorimeter system is by using the same particles for all individual sections. If these particles develop showers, then they can only be used to calibrate sections in which these showers are completely contained.**

Only in this way is the relationship between the deposited shower energy (in GeV) and the charge (in picoCoulombs) generated as a result established unambiguously. We have referred to this as the $B/A = 1$ method. The use of a beam of muons to intercalibrate the eighteen segments of the AMS-02 electromagnetic calorimeter (Sect. 9.3.5) definitely qualifies as a viable method in this respect. The mistake made in that case did not concern the calibration method itself, but the interpretation of the results.

An example of a good $B/A = 1$ method for calibrating calorimeters consisting of separate em and hadronic compartments is to determine the calibration constant of the em compartment with electrons of known energy and to intercalibrate the different longitudinal segments with a beam of muons. In non-compensating calorimeters, the energy of showering hadrons is not correctly reproduced in this way. However, that problem can be handled by applying an overall correction factor, which can be measured independently. An example of a calorimeter that was calibrated in this way is the CCFR Target calorimeter that operated in Fermilab's neutrino beam [16].

Another correct ($B/A = 1$) calibration method is to expose the individual longitudinal segments separately to an electron beam and derive the calibration constants from the signals recorded in that way. In many cases, constructional details prohibit such a procedure from being applied. Yet, this method was used for the calibration of the prototype modules of the CMS calorimeter system, results of which are shown in Figs. 7.3 and 7.6.

The CMS calorimeter system suffers from a particularly nasty complicating factor, namely the fact that the em and hadronic compartments have very different $e/h$ values. The em section of the barrel calorimeter is made of $PbWO_4$ crystals ($e/h \approx 2.5$), the hadronic section consists of brass plates interleaved with plastic scintillator ($e/h \approx 1.4$). For a systematic study of the hadronic performance of this calorimeter system, both compartments were calibrated with a 50 GeV electron beam.

Figure 7.6 shows that the response to pions, represented by the black dots, is very non-linear. This non-linearity is especially evident below 10 GeV, which is important since pions in this energy range carry a large fraction of the energy of jets at the LHC. More troublesome is the fact that the response depends on the starting point of the showers. The figure shows results for two event samples, selected on that basis: showers starting in the em section ($\triangle$) or in the hadronic section ($\triangledown$).

In CMS, the energy dependence of the response shown in Fig. 7.6 is the basis of the correction for the observed non-linearity; measured signals are simply multiplied with the inverse of the response value for that particular energy value. The use of this energy dependent overall correction factor restores hadronic signal linearity on average, even though the starting point dependence remains. If the starting point of the hadron shower can be determined, then the inverse of the $\triangle$ or $\triangledown$ curves from Fig. 7.6 could be used as the correction factor instead.

Just like in CMS, the em and hadronic sections of the ATLAS calorimeter system also have different $e/h$ values, but the differences are much smaller in that case. The consequences of these differences are qualitatively the same, but less dramatic than indicated in Figs. 7.3 and 7.6. Compensating calorimeter systems offer more calibration possibilities than ATLAS and CMS, because several other methods are equivalent to $B/A = 1$ in that case.

## 9.3.12   Validation

When all is said and done, it is important to check the correctness of the chosen calibration scheme with experimental data. For this, one needs to have sources of known experimental energy deposits. One such source is a particle of precisely known mass, whose decay products are detected by the calorimeter. This method offers excellent possibilities to validate the calibration of em calorimeters, because of the availability of a variety of particles that are abundantly produced in today's accelerator experiments and which cover a large mass (i.e., energy) range: $\pi^0$ (mass 135.0 MeV/$c^2$), $\eta$ (547.9 MeV/$c^2$), $J/\psi$ (3.097 GeV/$c^2$), $\Upsilon$ (9.460 GeV/$c^2$) and $Z^0$ (91.19 GeV/$c^2$) all decay into particles ($e^+e^-$ or $\gamma$ pairs) that develop electromagnetic showers. Examples of mass peaks reconstructed from these decay products are shown in Fig. 9.10 [17].

Unfortunately, there are no such clearcut calibration sources that can be used for hadron calorimeters. Yet, there are certainly possibilities. At low energies, one could use the decay $K_S^0 \rightarrow \pi^+\pi^-$ for this purpose and at high energies the hadronic decay modes of the intermediate vector bosons $W$ and $Z$. In the latter case, the problem is the QCD background, which makes it hard to extract a sample of boson decay events from the di-jet invariant mass distributions. The only experiment that has managed to do so was UA2 [18]. However, at the LHC one could take advantage of the high production rate of $t\bar{t}$ events, and select a much cleaner sample of hadronically decaying $W$s from the dominant decay mode $t \rightarrow Wb$.

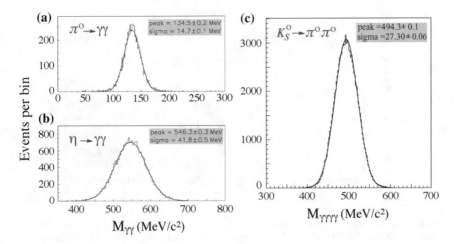

**Fig. 9.10** Invariant mass distributions of two $\gamma$s and four $\gamma$s detected in the same event by the KLOE em calorimeter. The mass peaks of the $\pi^0$ (**a**), $\eta$ (**b**) and $K_S^0$ (**c**) are located within 1% of their established values. From: Adinolfi, M. et al. (2002). *Nucl. Instr. and Meth.* **A482**, 364

Alternative methods to validate the hadronic calorimeter calibration include:

- A comparison between the momenta of isolated charged tracks and the corresponding energy deposits in the calorimeter.
- The use of events in colliders for which the energy deposited in the calorimeters is precisely known. As examples we can mention hadronically decaying $Z$ bosons produced at LEP and neutral current events in the $ep$ collisions at HERA, in which the hadronic four-vector was completely constrained.
- A method called "$\gamma$ + jet $p_T$ balancing", in which events are selected with one clear jet plus another object that develops an em shower and has no tracks associated to it. The transverse momentum of the jet is assumed to be equal to that of the "$\gamma$". This method has, for example, been used by CDF [19], and is also part of the CMS calibration procedure [20].

However, none of these alternative methods provides a check with the same level of rigorosity as the ones based on the reconstruction of a particle with a precisely known mass from its decay products.

## 9.4 Consequences of Miscalibration

As we have seen in the previous section, the calibration of a longitudinally segmented calorimeter system is far from trivial. It would, therefore, not be surprising at all if many of the calorimeter systems in current and past experiments were calibrated incorrectly. However, it is in practice not easy to find experimental proof of that.

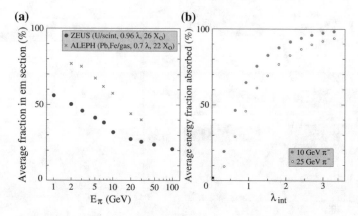

**Fig. 9.11** Energy sharing between the electromagnetic and hadronic calorimeter sections for pions detected with the ALEPH and ZEUS calorimeters [2]. Diagram **a** shows the fraction of the energy deposited in the em calorimeter section, as a function of energy. Experimental data from [11, 21]. Diagram **b** shows the energy fraction deposited by 10 and 25 GeV pions in iron absorber as a function of the absorber thickness. Experimental data from [22]

One example of a case in which miscalibration manifests itself beyond a shadow of a doubt is when particles of precisely known mass (e.g., the $J/\psi$ meson or the $Z^0$ boson) are reconstructed with a significantly different mass value on the basis of their calorimetrically measured decay products.

Sometimes, there may also be indirect indications of problems caused by miscalibration. In this section, we discuss one example of this in some detail.

Figure 9.11 contains data from pions developing showers in the ALEPH [11] and ZEUS [21] calorimeters. The ALEPH calorimeter consisted of iron absorber, read out with wire chambers; ZEUS operated a uranium/plastic-scintillator device. In Fig. 9.11a, the average fraction of the pion energy deposited in the em calorimeter compartment is plotted as a function of the pion energy for both calorimeter systems.

The pions deposited a considerable fraction of their energy in the em calorimeter section. At 10 GeV, the average fraction was measured to be ∼30% for ZEUS and ∼60% for ALEPH. The differences between these two sets of results were very large. Moreover, they had the wrong sign. The ALEPH em calorimeter compartment was *thinner* than the ZEUS one (0.7 vs. 0.96 nuclear interaction lengths). Based on measurements of the hadronic shower containment as a function of the absorber thickness, also done by ALEPH [22], one would expect the energy fraction deposited by pions in ALEPH's em calorimeter section to be 40% *smaller* than for ZEUS, instead of a factor two larger (Fig. 9.11b). Given this large discrepancy (a factor of 2.5 at 10 GeV), the conclusion that one of the two experiments (or maybe both) mismeasured the energy deposited by pions in the electromagnetic calorimeter compartment by a considerable factor is hard to avoid.

The above-mentioned discrepancy can be understood by analyzing the calibration methods that were used to obtain these data. Contrary to ZEUS, the ALEPH calorimeter was strongly non-compensating, with $e/\pi$ signal ratios ranging from 1.7 at 2 GeV to 1.3 at 30 GeV. This means that pions produced signals that were considerably smaller than those from electrons at the same energy. To get to the correct energy, the calibration constants ALEPH used for pions were therefore considerably larger than for electrons.

The philosophy of the ALEPH calibration method, which was essentially the $(B/A < 1)$ one discussed in Sect. 9.3.8, was to achieve a situation in which

1. pions and electrons of the same energy produce, *on average*, the same signals, and
2. the *average* pion signal is proportional to the energy (i.e., hadron linearity).

The price that had to be paid for reaching that goal was a considerable mismeasurement of the energy of particles or jets whose signals did *not* resemble those of the average pion. By overestimating the energy deposited in the em calorimeter section by more than a factor of two, the energy of particles or jets that happened to deposit a considerable fraction of their energy in this calorimeter section was similarly overestimated. And the energy of particles that penetrated the em section without undergoing a strong interaction was systematically underestimated by a considerable factor.

The most serious problems resulting from this calibration strategy may be expected for jets. First, since jets contain some fraction of $\pi^0$s, which *always* deposit their *entire* energy in the em calorimeter section, the *average* energy sharing between the two sections may be quite different for charged pions and for jets of the same energy. The latter deposit, on average, a larger fraction of their energy in the em calorimeter section. As a result, the jet energy is, on average, overestimated.

Second, the jet energy measurement is strongly correlated with the jet topology. If the leading particle in the fragmentation process that generated the jet was a $\pi^0$, then all its energy would be deposited in the em calorimeter section. If it were charged, then most of its energy would go to the hadronic section. The ALEPH detector would measure a very different value for the energy of these two jets. Assuming that the non-leading jet particles would generate the same signals in both cases, the difference between the energies measured for these two jets could be as much as 35% [23]. The signal distribution for mono-energetic jets detected with the ALEPH calorimeter would thus be considerably broader than the signal distribution for mono-energetic pions of the same energy.

Possible evidence for the problems that this calibration scheme may have caused for jet detection may be derived from Fig. 9.12. This figure shows the total energy distribution for hadronically decaying $Z^0$ particles measured with the ALEPH calorimeter system, and may thus be considered a jet measurement at 90 GeV.

Of course, some of the jets initiated by $b$ and $c$ quarks may have contained neutrinos and muons which were not (completely) absorbed by the calorimeters. Also, some energy may have leaked out due to the fact that the detector did not cover the complete $4\pi$ solid angle. Such phenomena cause a low-energy tail in the signal distribution, which can actually be observed in Fig. 9.12.

**Fig. 9.12** The total energy
distribution of hadronically
decaying $Z^0$ particles,
measured with the
calorimeters of the ALEPH
experiment at LEP. The
black data points were used
for the fit described in the
text. From: ALEPH
Collaboration, 1991, private
communication

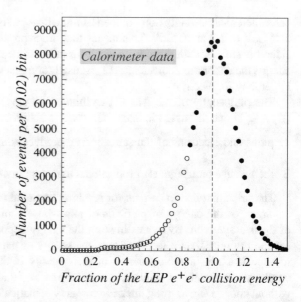

When this tail was ignored and the distribution in the area $E/E_{LEP} = 0.85 - 1.5$
(the black experimental points in Fig. 9.12) was fitted to a Gaussian function, a
fractional width $\sigma$/mean of 14.3% was obtained. This may be compared with a
width of 8.8% that one would have expected on the basis of the energy resolution
quoted for single hadrons ($\sigma/E = 84\%/\sqrt{E}$ [11]).

The fact that not one, but a multitude of hadrons (an unknown and fluctuating
mixture of $\pi^0$s and other particles) were measured simultaneously by the calorimeter,
introduced thus an extra contribution of $\sim$11% (i.e., $\sqrt{14.3^2 - 8.8^2}$) to the energy
resolution in the case of 90 GeV jets.

It should be emphasized that this discussion is by no means intended to criti-
cize ALEPH. This example was chosen because of its educational value. It clearly
illustrates the problematic consequences of a wrong calibration strategy.

In the LEP studies of $Z^0$ decay, hadronic calorimetry played only a very modest
role. The ALEPH experiment was equipped with very powerful tracking capabil-
ities and offered, therefore, excellent alternatives to calorimetry for studying the
$Z^0$ physics. The ALEPH Collaboration demonstrated that the distribution shown in
Fig. 9.12 could be considerably improved when the calorimetric information on the
charged hadrons was replaced by the measured track momenta [24]. If, in addition,
the event sample was limited to $Z^0$s that produced jets in the central detector region,
then the resolution improved from 14.3% to 6.9%.

However, it should also be emphasized that the very clean nature of the $e^+e^- \rightarrow$
$Z^0$ events, which made such improvements possible, is by no means representative
for today's experiments in particle physics, which increasingly rely on excellent and
correctly calibrated calorimeter systems for their physics analyses.

## 9.5 Off-Line Compensation

In the previous sections we have shown several examples of problems that may arise when a calorimeter system consisting of two longitudinal segments is incorrectly calibrated. Similar problems occur when the calorimeter consists of three, four, five or even more longitudinal segments. Also for such systems, the only correct way to intercalibrate the various segments is with particles that behave, on average, identically in each and every segment.

Only in this way is the energy deposited in the active calorimeter material measured unambiguously. Alternative calibration methods, designed to correct the undesirable effects of $e/h \neq 1$ by means of weighting factors, optimized for an "average shower", tend to amplify these effects for non-average events.

Sometimes, a longitudinal subdivision into many segments is applied in the hope of achieving the advantages offered by an intrinsically compensating calorimeter, for example the superior hadronic energy resolution. However, this cannot be achieved through general weighting factors, since resolutions are determined by event-to-event fluctuations, not by mean values. If and only if one succeeds in measuring the em shower content *event by event* can one reduce the effect of event-to-event fluctuations in the em shower content, which tend to dominate the hadronic energy resolution of non-compensating calorimeters. The dual-readout method discussed in Sect. 8.3 successfully achieves that goal. Prior to that, attempts were developed to use differences between the profiles of the em and non-em components of hadron showers to that end. This method, that became known as *off-line compensation*, was pioneered by the WA1 Collaboration [25].

The H1 Collaboration [26] applied a calibration method in which large local energy deposits were selectively suppressed. This also eliminated some of the consequences of event-to-event fluctuations in the em content of hadronic showers. Their LAr calorimeter consisted of an em compartment subdivided into five longitudinal segments and a hadronic compartment subdivided into six such segments. The total energy was calculated as

$$E = C_e^w \sum_{i=1}^{5} Q_i (1 - \eta_e Q_i) + C_h^w \sum_{i=1}^{6} Q_i (1 - \eta_h Q_i) \tag{9.5}$$

where $Q_i$ represented the charges in the $(5 + 6)$ individual longitudinal segments, $C_e^w$ and $C_h^w$ set the overall scale, and the factors $(1 - \eta_e Q_i)$ and $(1 - \eta_h Q_i)$ were required to exceed a minimum value ($\delta$). The latter requirement suppressed large energy deposits in individual cells.

This method thus involved five parameters ($C_e^w$, $C_h^w$, $\eta_e$, $\eta_h$, $\delta$). With the exception of $\delta$, these parameters were found to be energy dependent, just as in the calibration of the ATLAS ECAL (Fig. 9.6).

Such calibration methods improved the resolution for single pions from beams with known energy, at least for energies in excess of 30 GeV. Also, the line shape became more symmetric as a result of recognizing and correcting for the anomalously

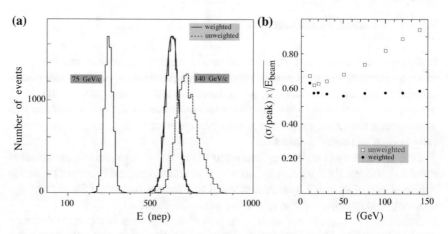

**Fig. 9.13** WA1 results on offline compensation. The signal distributions for 140 GeV pions (**a**) and the hadronic energy resolution as a function of energy (**b**), before and after the weighting procedure described in the text was applied to the experimental data. From: Abramowicz, H. et al. (1981). *Nucl. Instr. and Meth.* **180**, 429

large concentrations of energy deposit responsible for the high-end tails of pionic signal distributions in a non-compensating calorimeter.

The experimental data (Fig. 9.13) showed that procedures of this type worked especially well at high energies, where $\pi^0$s and non-em forms of energy deposit led to distinctly different signals in individual longitudinal segments of the detectors. At low energies, where these differences were much less spectacular, the improvements in energy resolution were found to be marginal, at best. Also the hadronic signal non-linearity did not disappear at energies below 30 GeV. Neither WA1 nor H1 have demonstrated any beneficial effects of these "off-line compensation" methods for jets, or more generally for a situation in which energy is deposited in the calorimeter system by a collection of particles with unknown composition and energies.

The ATLAS experiment has been inspired by the technique developed by H1 to correct their hadronic energy measurements for the effects of non-compensation. The ATLAS barrel calorimeter consists of a LAr ECAL plus an iron/plastic-scintillator HCAL, each of which is subdivided into three longitudinal compartments. In addition, there is a presampler intended to recover energy lost in the material upstream of the calorimeter system. In the beam tests on the basis of which their "cell weighting method" was developed, the hadronic calorimeter consisted of four longitudinal compartments. In addition, a "mid sampler" was installed to account for energy lost in the cryostat walls separating both calorimeter systems [27].

Also in their approach, the reconstruction of the energy of the showering hadron relies on upwards corrections of relatively small signals. The signals from cells that are relatively small compared to those from cells in which em shower components deposit energy are given a weight intended to equalize the response of those two types of cells. The total energy is determined with a formula similar to (9.5):

$$E = \sum_{\text{em cells}} W_{\text{em}}(E_{\text{cell}}, E_{\text{beam}}) E_{\text{cell}} + \sum_{\text{had cells}} W_{\text{had}}(E_{\text{cell}}, E_{\text{beam}}) E_{\text{cell}} + E_{\text{cryo}} \quad (9.6)$$

The weight factors applied to the signals from each individual calorimeter cell are a function of the compartment in which the calorimeter cell is located and of the beam energy:

$$W_{\text{em}} = A_E + B_E / E_{\text{cell}} \qquad (9.7)$$
$$W_{\text{had}} = A_H + B_H / E_{\text{cell}} \qquad (9.8)$$

The parameters $A_E$, $B_E$, $A_H$ and $B_H$ were taken from a fit and all depend on the hadron energy. Since finding the hadron energy was the purpose of the entire exercise, the parameter values had to be determined in an iterative process. The term $E_{\text{cryo}}$ is intended to account for the loss of energy in the cryostat and is proportional to the geometric mean of the energy lost in the last em compartment and the first hadronic one. It turned out that $E_{\text{cryo}}$ defined in this way was nicely correlated with the energy detected in the "mid sampler", so that there was no need to include the latter device in the design of the final detector.

As stated before, procedures of this type may be applied with some success in beam tests, where the properties of the particle that generates the signals are precisely known. However, the benefits in the messy environment of a high-luminosity hadron collider are not clear, and in any case remain to be demonstrated.

## 9.6 Conclusions

We have distinguished longitudinally unsegmented and segmented calorimeter systems. The first category took less than one page to discuss, the rest of the chapter was almost entirely spent on the second category.

Calibrating longitudinally segmented calorimeter systems is very tricky indeed. This is caused by the fact that the calorimeter response depends on the type of particle, on its energy and on the age of the shower it developed. Especially in jets, the various effects that play a role interfere in ways that are impossible to disentangle.

We have discussed four examples of strategies that are used in practice to calibrate a calorimeter system consisting of separate longitudinal compartments:

1. In the first strategy, the calibration constants were chosen such as to optimize the energy resolution for particles that deposited their energy in several calorimeter compartments (Sects. 9.3.2–9.3.6).
2. In the second strategy, the energy scale was set by particles that deposited their entire energy in one of the two compartments: electrons for the em compartment, pions that penetrated the em compartment without interacting for the hadronic compartment (Sect. 9.3.7).

3. The third strategy aimed for hadronic signal linearity in combination with reconstruction of the beam energy for electrons and pions simultaneously. This was done by increasing the signals from the em calorimeter compartment for hadron showers by a weighting factor (Sect. 9.3.8). A similar strategy was used to make the hadron response independent of the starting point of the showers (Sect. 9.3.9).
4. The fourth strategy aimed to achieve equal responses for electrons and pions by multiplying the signals from the ($8X_0$ thick) first calorimeter segment with zero (Sect. 9.3.10).

All four strategies looked, at first sight, reasonable and logical. They were all designed to achieve a specific beneficial result: optimization of the energy resolution (method 1), signal linearity for hadrons (method 3), equalization of the response to electrons and pions (method 4). Yet all four methods turned out to be fundamentally flawed at more or less severe levels. In each case, the benefit for which the calibration method was designed was offset by the introduction of new, sometimes very severe, problems such as

- Signal non-linearity for em showers (1,4)
- Systematic mismeasurement of the energy of early (2) or late (3) showering particles
- Systematic mismeasurement of the energy of hadrons that penetrate the em calorimeter compartment without starting a shower (3)
- Systematic mismeasurement of the energy of jets (2,3,4)
- A general degradation of the energy resolution and the line shape (3,4)
- Systematic mismeasurement of the energy of $\gamma$s and $\pi^0$s (1,4)

The underlying reason for all these problems is the fact that the mentioned calibration strategies only work *on average*. The envisaged goals are only achieved for a particular subset of events, for example the pions that penetrate the em compartment without interacting (method 2), or electrons of a given energy (method 1), or pions that deposit $44 \pm 1\%$ of their energy in the em calorimeter compartment (method 3). At the same time, new problems are introduced for events that do not resemble the average for which the calibration method was designed. Manipulating calibration constants may thus seem to work out fine for one particular subset of events, but may have very negative consequences for other, possibly more relevant, subsets.

The only way to avoid such problems is to calibrate the individual sections of a longitudinally segmented calorimeter system in exactly the same way ($B/A = 1$). If the individual compartments can be separated and are deep enough to contain electron showers, then one may use an electron beam for this purpose. If that is not possible, then these sections may be intercalibrated with muons traversing the entire depth of the calorimeter. Only in this way is the relationship between the deposited shower energy and the resulting signal established unambiguously, identically for all calorimeter sections.

In doing so, the conditions that exist automatically in longitudinally *unsegmented* calorimeters and that make these devices trivial to calibrate are reproduced. This leads us to the conclusion that longitudinal segmentation, in general, *does not serve any purpose* with regards to the precision with which the energy of showering particles can be measured in a calorimeter.

Only if the segmentation makes it possible to determine the em shower content **event by event** could some benefit be expected. However, even the feasibility of this is questionable in environments other than a testbeam where particles of precisely known energy and type are delivered to the detector.

The above conclusion may be illustrated by considering a perfectly compensating calorimeter, i.e., a calorimeter in which event-to-event fluctuations in the em shower content do not contribute to the energy resolution. If such a calorimeter were longitudinally segmented, then the extra information on the shower development would not allow one to improve the energy resolution. In fact, one could only deteriorate the excellent resolution of the unsegmented device by calibrating the segments in a way that differs from the only correct one: $B/A = 1$.

We have noticed that there is a general belief that collecting more information about the absorption process inside calorimeters, which is the goal of having multiple longitudinal segments, must necessarily lead to achieving better performance characteristics. We think that this is an illusion, for the same reason as it is an illusion to think that by measuring the four-vectors of all the individual molecules of the gas in a certain container one would be able to make a more precise determination of the temperature inside that container. It would just have the effect of complicating the issues to a degree that would make it almost impossible to avoid making mistakes. Something very similar is happening here. Simple is better, and money for readout is much better spent on lateral segmentation (granularity!) than on an increased number of longitudinal compartments.

A fine lateral segmentation could also be very helpful for the identification of electrons and $\gamma$s, one of the other issues that is often used as an argument in favor of longitudinal segmentation. This is further discussed in the context of applications of calorimeter information for particle identification (Chap. 11).

# References

1. Sefkow, F., et al.: Rev. Mod. Phys. **88**, 015003 (2016)
2. Wigmans, R.: Calorimetry—Energy Measurement in Particle Physics, 2nd edn. International Series of Monographs on Physics, vol. 168. Oxford University Press, Oxford (2017)
3. Åkesson, T., et al.: Nucl. Instrum. Methods **A262**, 243 (1987)
4. Wigmans, R., Zeyrek, M.: Nucl. Instrum. Methods **A485**, 385 (2002)
5. Aharouche, M., et al.: Nucl. Instrum. Methods **A568**, 601 (2006)
6. Aabouid, M., et al.: J. Instrum. **14**, P03017 (2019)
7. Cervelli, F., et al.: Nucl. Instrum. Methods **A490**, 132 (2002)
8. Adloff, C., et al.: Nucl. Instrum. Methods **A714**, 147 (2013)
9. Braunschweig, W., et al.: Nucl. Instrum. Methods **A265**, 419 (1988)
10. Bertolucci, S., et al.: Nucl. Instrum. Methods **A267**, 301 (1988)
11. Bagliesi, G., et al.: Nucl. Instrum. Methods **A286**, 61 (1990)
12. Ganel, O., Wigmans, R.: Nucl. Instrum. Methods **A409**, 621 (1998)
13. Albrow, M., et al.: Nucl. Instrum. Methods **A487**, 381 (2002)

14. Ferrando, A., et al.: Nucl. Instrum. Methods **A390**, 63 (1997)
15. Akchurin, N., et al.: Nucl. Instrum. Methods **A400**, 267 (1997)
16. Sakumoto, W.K., et al.: Nucl. Instrum. Methods **A294**, 179 (1990)
17. Adinolfi, M., et al.: Nucl. Instrum. Methods **A482**, 364 (2002)
18. Alitti, J., et al.: Z. Phys. C **49**, 17 (1991)
19. Bocci, A., et al.: Int. J. Mod. Phys. A **16**(suppl. 1A), 255 (2001)
20. CMS Collaboration: Note CMS-PAS-PFT-09-001 (2009)
21. Behrens, U., et al.: Nucl. Instrum. Methods **A289**, 115 (1990)
22. Catanesi, M.G., et al.: Nucl. Instrum. Methods **A292**, 97 (1990)
23. Wigmans, R.: Ann. Rev. Nucl. Part. Sci. **41**, 133 (1991)
24. Buskulic, D., et al.: Nucl. Instrum. Methods **A360**, 481 (1995)
25. Abramowicz, H., et al.: Nucl. Instrum. Methods **180**, 429 (1981)
26. Braunschweig, W., et al.: Nucl. Instrum. Methods **A275**, 246 (1989)
27. Akhmadaliev, S., et al.: Nucl. Instrum. Methods **A449**, 461 (2000)

# Chapter 10
# Operational Challenges

The calorimeters used in experiments at colliders do not operate in the ideal conditions that are typical for beam tests of prototype modules. Instead, they face a challenging environment that tends to make the extraction of the envisaged physics information, and sometimes the very operation of the instrument, problematic. This is increasingly true for experiments at each new generation of colliders, since the cross sections of the interesting processes and the background obscuring these processes increases with the center-of-mass energy of the collisions.

In this chapter, we discuss three sources of problems that play a role in this context:

1. Magnetic fields
2. Radiation damage
3. High luminosity

## 10.1 Operation in a Magnetic Field

In many experiments, the calorimeters operate in a magnetic field that serves to determine the momenta of individual charged particles upstream, and sometimes downstream, of the calorimeter system. This has a variety of practical consequences. It limits the choice of materials that can be used in the construction of the calorimeter system, it imposes limitations on the readout technology that can be used, and it affects the calorimeter signals.

### 10.1.1 Construction Materials

The use of magnetic materials, such as iron or nickel, should be avoided in the construction of the calorimeters. The forces exerted on these materials when the magnetic field is switched on could cause severe damage to the experimental equipment. But

© Springer Nature Switzerland AG 2019
M. Livan and R. Wigmans, *Calorimetry for Collider Physics, an Introduction*,
UNITEXT for Physics, https://doi.org/10.1007/978-3-030-23653-3_10

even if everything were so rigidly constructed that nothing physically moves when the field is switched on, the iron could severely distort the magnetic field itself, causing problems for the reconstruction of the trajectories of charged particles.

## 10.1.2  Signal Readout in a Magnetic Field

Operation in a magnetic field also imposes limitations on the readout technology that can be used. If the calorimeter produces light signals, then it is often not possible to use photomultiplier tubes as light detecting elements.

Most PMTs are so sensitive to magnetic fields that already fields as weak as the Earth's magnetic field ($\sim 5 \cdot 10^{-5}$ T) may have a significant effect on the gain. It can be easily checked whether or not this is the case for a certain setup by changing the orientation, since only the non-axial component of the field affects the trajectories of the photoelectrons.

To avoid such effects, PMTs are usually equipped with shields made of material with a high magnetic susceptibility ("$\mu$-metal"), which surround the sensitive areas (mainly the photocathode). Some PMTs have been specially designed to be able to operate in magnetic fields. This is mainly achieved through a very compact dynode structure, as in *close proximity focusing* tubes [1]. Also compact structures with only one or two dynodes (known as *vacuum phototriodes* or *phototetrodes*) have been demonstrated to be capable of operating in moderate magnetic fields, up to $\sim 1$ T. Both OPAL [2] and DELPHI [3] have used such devices to read out their lead-glass em calorimeters. CMS uses phototriodes for reading out the PbWO$_4$ crystals of the endcap sections of their electromagnetic calorimeter [4], where the (solenoidal) magnetic field is almost axial. These particular devices have an anode of very fine copper mesh (10 $\mu$m pitch), which allows them to operate with $<10\%$ gain loss in the 4 T magnetic field of CMS [5].

Alternative light detectors capable of operating in (strong) magnetic fields include avalanche photodiodes (APDs) and silicon photomultipliers (SiPMs). However, each of these alternative solutions may have disadvantages, compared with PMTs. For example, these devices may be very unstable under small changes in temperature or bias voltage. APDs are also very sensitive to charged shower particles crossing them, which may lead to catastrophic effects (Sect. 4.3), and SiPMs are intrinsically non-linear devices (Sect. 4.2.2).

The effects of magnetic fields on detectors depend strongly on the orientation of the magnetic field with respect to the electric field that accelerates the (photo)electrons towards the collecting electrode. The effects are of course minimal when $\mathbf{B} \parallel \mathbf{E}$. They reach a maximum when $\mathbf{B} \perp \mathbf{E}$, because the slow (photo)electrons are bent away from the direction they should follow to produce a signal.

One way to deal effectively with the possible problems caused by magnetic fields in light detectors is to transport the light signals from their source to an area where the magnetic field is so low that it does not prohibit readout with PMTs. Clear plastic fibers allow one to do so with minimal light losses. Tests done in the context of the development of the CDF Plug Upgrade calorimeter by a group at Fermilab have shown that the light losses incurred in coupling fibers in which the detector signals were generated to such clear fibers could be limited to 10–15%, with good reproducibility [6].

In calorimeters using liquid or gaseous dielectrics, magnetic fields cause limitations for the type of electronic circuits that can be used. For example, voltage or current transformers do not work properly when placed in an external magnetic field. In gas calorimeters based on drift chambers as active elements, the path and the velocity vector of the drifting electrons may be altered by the Lorentz force. In that case, a precise knowledge of the magnetic field map throughout the detector is crucial for determining the modified relationships between drift time and position.

### 10.1.3  Effects on the Calorimeter Signals

Apart from the effects that magnetic fields have on the functioning of detectors and electronics that handle calorimeter signals, they may also affect these signals themselves. We mention two examples.

*Increased Light Yield*

The light yield of some plastic scintillators changes when placed in a magnetic field. Typically, the light yield increases, by some 5–10%, for magnetic fields up to 3 T [7–11]. For larger fields, no further increase was observed by Bertoldi and coworkers [11], who tested a variety of different scintillators and wavelength shifters for fields up to 20 T. The same authors also reported that the increased light yield was only observed when the scintillators were excited with ionizing particles. In particular, no significant changes in the light yield were observed when the excitation was performed with UV light.

This is a strong indication that the effects are due to an increased production of UV light in the excitation of the polymer base material (polystyrene or polyvinyl toluene), since the excitation of the scintillating fluors with UV light was not affected by the magnetic field.

The light yield is not a simple function of the field strength, and there is no experimental indication that the effects depend on the orientation of the magnetic field or on the nature of the ionizing radiation.

*Effects on Shower Profiles*

Magnetic fields may also affect the shower profiles, since the paths of the charged shower particles are subject to the Lorentz force. This may have consequences for the early, non-isotropic shower component. These consequences depend on the strength and the orientation of the magnetic field. If the field is perpendicular to the shower axis, then the lateral shower profile will be broadened. A large fraction of the signals from electromagnetic showers comes from soft electrons, produced in Compton scattering and in photoelectric processes (Sect. 3.4). For example, in a copper-based calorimeter, more than half of the shower energy is deposited by electrons softer than 4 MeV (Fig. 3.9). The trajectory of these soft electrons is very sensitive to magnetic fields, since they are subject to the Lorentz force. For example, the radius of curvature of an electron with a momentum of 4 MeV/$c$ in a 2 T magnetic field oriented perpendicular to its direction of motion is only about 6 mm.

These soft electrons only contribute to the signals from sampling calorimeters if they are produced in a very thin boundary layer. The range of 4 MeV electrons in copper is only 3 mm (Fig. 3.12a). If a 4 MeV electron escaped from the copper and traversed a 5 mm thick plastic scintillator layer perpendicularly, then it would lose typically 1 MeV. However, if the calorimeter were placed in a strong magnetic field oriented parallel to the sampling layers, then this escaping electron would describe a curved trajectory and lose a considerably larger fraction of its energy in the scintillator. Depending on the gap width between the absorber plates and the strength of the magnetic field, it could even reverse direction and deposit its *entire* energy in the scintillator plate, in a way similar to the "cork-screws" that were the signature of electrons in bubble-chamber pictures (Fig. 10.1a).

This effect results in an increased calorimeter response. It increases with the strength of the magnetic field, since the fraction of shower particles trapped in the gaps between the absorber plates increases with the field strength. It would not play a role for magnetic fields oriented perpendicular to the sampling layers.

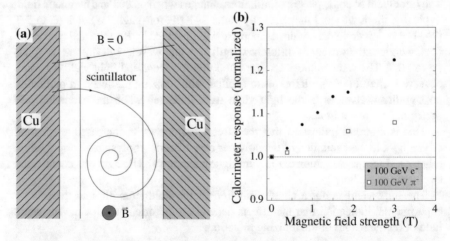

**Fig. 10.1** Trajectories of few-MeV electrons contributing to the signals of a sampling calorimeter in the absence and presence of a magnetic field oriented parallel to the sampling layers, pointing into the plane of the figure (**a**). The relative increase in the response of the CMS copper/plastic-scintillator calorimeter, as a function of the strength of a magnetic field oriented parallel to the sampling layers (**b**). The response dependence is given for showers induced by 100 GeV electrons and by 100 GeV pions. In order to eliminate the effects on the specific light yield of the scintillator, all responses have been normalized to those for muons traversing the calorimeter in the same direction as the showering particles. From: Kunori, S. (1997). *Proc. 7th Int. Conf. on Calorimetry in High Energy Physics*, Tucson, Arizona (Singapore: World Scientific, Singapore), p. 224

The described effects were observed by the CMS Collaboration in prototype tests of their hadron calorimeter [12]. Since the copper/plastic-scintillator CMS calorimeter has to operate in a very strong magnetic field, the effects of this field on the calorimeter performance were studied in great detail. Some results of these studies are shown in Fig. 10.1b, where the relative increase in the calorimeter response is plotted as a function of the magnetic field strength, for showers induced by 100 GeV electrons and 100 GeV pions. The field was oriented parallel to the sampling layers in these studies. In order to eliminate the "scintillator brightening" effects discussed in Sect. 10.1.2, the calorimeter responses to these particles were normalized to those for muons that traversed the calorimeter in the same direction. The em response was measured to increase by about 20% when the field strength reached 3T, the maximum value for which measurements were performed in these studies.

The effect of the magnetic field on the hadronic response was clearly smaller, commensurate with what should be expected if only the em shower component was affected by the field. This makes perfect sense, since hadronic shower particles with similar sensitivity to the Lorentz force as the electrons from the example discussed above are extremely non-relativistic and would thus deposit their entire kinetic energy in the scintillator, with or without a magnetic field. Therefore, the field does not affect the response to the non-em shower component. Since only the em shower component is affected, the $e/h$ value of the calorimeter is thus increased when it operates in a magnetic field.

The CMS tests showed no measurable effect on the calorimeter response (over and above the scintillator brightening discussed above), if the magnetic field was oriented *perpendicular* to the sampling layers, i.e., parallel to the shower axes.

## 10.2   Radiation Damage

As the luminosities at colliding-beam machines have increased (e.g., to compensate for the fact that the cross sections for the interesting processes rapidly decrease as $\sqrt{s}$ increases), radiation damage of crucial detector components has become more and more a point of concern. At hadron colliders, scintillation based calorimeters have become less favored because of the sensitivity of active media based on light production to ionizing radiation. Even CMS, which choose lead tungstate crystals based on the (advocated and) supposed radiation hardness, had to face the fact that the endcap region of their calorimeter system is fast becoming unusable, after having received only a small fraction of the envisaged total integrated luminosity of the Large Hadron Collider.

This is illustrated in Fig. 10.2 [13]. The left diagram shows the effect of ionizing radiation on the light transmission, as a function of wavelength. Especially the short wavelength range, which is crucial for the transmission of the scintillation light generated by the crystals, is affected. The dominating source of damage are the hadrons. Unlike the effects caused by photons, the damage induced by hadrons cannot be undone by annealing at room temperature. The right diagram in Fig. 10.2

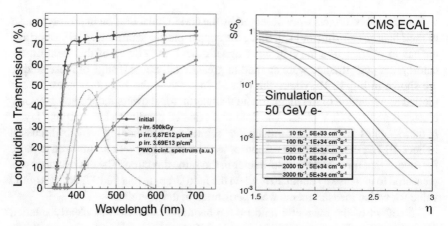

**Fig. 10.2** Effects of radiation damage on the performance of the CMS em calorimeter. The left diagram shows the light transmission in the PbWO$_4$ crystals after irradiation with $\gamma$s and protons. The overlaid black dotted line represents the lead tungstate light emission spectrum. The right diagram shows the (simulated) effect on the scintillation signals from 50 GeV showers in the crystals as a function of the pseudorapidity, for various values of the integrated luminosity. From: CMS Collaboration (2015). Technical proposal for the phase-II upgrade of the Compact Muon Solenoid, CERN-LHCC-2015-10

shows the reduction of the PbWO$_4$ signals as a function of the pseudorapidity at which the crystals are installed in the endcap region. At the present time, the signals are already down by a factor of two in the regions close to the beam pipe, and the effects are projected to become much worse over time (up to a reduction by a factor 1000 after the envisaged total integrated luminosity).

In order to better deal with high radiation levels, modern calorimeters are based on liquid active material, such as liquid argon, which can be relatively easily replaced, or on intrinsically (more) radiation hard materials, such as silicon. The CMS Collaboration has chosen a silicon based sampling calorimeter as a replacement for the crystals in the endcap section of their detector. Contrary to light based systems, radiation damage by ionizing particles is not the main problem in silicon. Rather, neutrons that scatter off the silicon nuclei and alter the lattice structure of the semiconductor material are the main source of concern in this case.

In the environment that the CMS endcap calorimeter has to face in the high-luminosity LHC, it is expected that the detectors will have to deal with a total neutron fluence of up to $10^{16}$/cm$^2$ close to the beam pipe, for the integrated total luminosity of 3000 fb$^{-1}$. Figure 10.3 shows the effects of exposure to neutrons on the signals from a number of different silicon detectors, as a function of the fluence. At the maximum expected rates, the signals are reduced by a factor of about four. These results are for silicon sensors with a thickness of 200 $\mu$m. Measurements have shown that the relative signal reduction decreases with the thickness of the depletion layer. For this reason, CMS envisages using thinner sensors near the beam pipe [13].

Not only the active calorimeter layers are subject to the effects of radiation. Damage of silicon based electronics embedded in the absorber structure is a point of

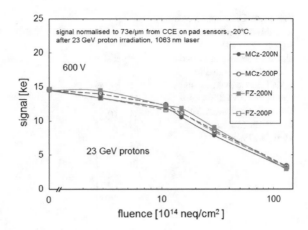

**Fig. 10.3** Effects of exposure to neutrons on the signals from different types of 200 μm thick silicon detectors, as a function of the total neutron fluence. The signals are expressed in kilo-electrons, the neutron fluence in 1 MeV neutron equivalent per cm². From: CMS Collaboration (2015). Technical proposal for the phase-II upgrade of the Compact Muon Solenoid, CERN-LHCC-2015-10

concern as well. As an example, I mention the case of the ATLAS Liquid Argon Calorimeter [14]. In preparation for the increased LHC luminosities foreseen for the future, ATLAS will replace the ASICs[1] that handle the calorimeter signals by more radiation hard ones. ASICs based on IBM's 130 nm CMOS (8RF) technology meet the requirements in that respect. The new ASICs will also be better adapted to handle the trigger rates expected at the higher luminosities. The analog on-detector Level-1 pipeline will be replaced by a system in which all calorimeter signals are digitized at 40 MHz and sent to the off-detector front-end electronics. This approach is expected to remove all constraints imposed by the calorimeter readout on the ATLAS trigger system.

## 10.3  Pileup

Radiation damage is not the only problem LHC experiments face as a result of the increasing luminosity. At the present time, ATLAS and CMS have to deal with, on average, 55 events per 25 ns bunch crossing (at a luminosity of $2 \cdot 10^{34}$ cm$^{-2}$s$^{-1}$). This rate is increasing proportionally with the luminosity and in the High Luminosity LHC era (after 2024), one will have to cope with a luminosity of $7 \cdot 10^{34}$ cm$^{-2}$s$^{-1}$, at which point interesting events accompanied by more than 200 "pileup" events will be no exception. Of course, the overwhelming majority of these "underlying events" are uninteresting, and involve relatively few high-$p_\perp$ particles that contribute to the detector signals. Yet, pileup induced background is expected to be a factor that seriously deteriorates the detector performance. High-precision timing is considered one of very few options to mitigate these effects, and this has given rise to a number of dedicated experimental studies [15–17].

---

[1] Application-Specific Integrated Circuit.

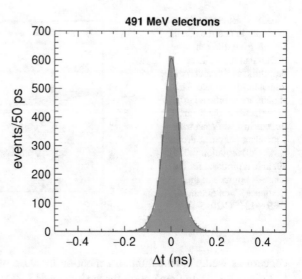

**Fig. 10.4** Distribution of the time difference between the signals from 491 MeV electrons measured in two MCPs, one operated in *i*-MCP mode and the other one in PMT-MCP mode. From: Barnyakov, A.Yu. et al. (2018). *Nucl. Instr. and Meth.* **A879**, 6

The elapsed time for an LHC bunch crossing has an rms spread of 170 picoseconds, which means that the 50–100 ps time resolution commonly achieved in the time-of-flight systems used for particle identification purposes is not adequate for solving this problem. One expects to need time resolutions of at least 20–30 ps to make a significant difference in this respect. A major complicating factor is that this performance has to be achieved in a very-high-rate environment. The approach followed by the mentioned R&D projects focuses on instantaneous light signals, such as those produced by the Čerenkov mechanism, combined with ultrafast photo detectors, such as Avalanche Photo Diodes (APDs) [15], Microchannel Plates (MCPs) [16], or micromegas [17].

Figure 10.4 shows results recently obtained with MCPs. These detectors are either operated in the standard PMT-MCP mode, or in the *i*-mode, in which the photocathode is removed and the signal is produced by secondary emission of electrons from the MCP layers crossed by the ionizing particles [18]. The figure shows the distribution of the time difference between signals from 491 MeV electrons measured in two MCPs, one operated in *i*-mode and the other in PMT-MCP mode. This Gaussian distribution has a width ($\sigma$) of 17±2 ps. A time resolution of 75 ps was reported *for single photoelectrons* by [17]. These are encouraging developments, but there is of course still a long way to go before systems capable of assigning signals to different events occurring in the same bunch crossing will be available for implementation in the extremely high-rate environment of LHC-type experiments.

In a more traditional approach to the pileup problem, it is treated as an additional source of electronic noise. Using a series of signal samples collected at intervals of 25 ns, the properties of the true signal are estimated on the basis of a variance minimization of the noise covariance matrix, a method known as optimal filtering [19]. The bipolar pulse shaping, introduced by Radeka [20], is crucial for the success of this method with the LAr signals from the ATLAS calorimeters. This method

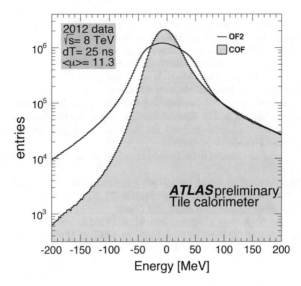

**Fig. 10.5** Cell energy distribution reconstructed by the Constrained Optimal Filter (COF) and by the Optimal Filtering (OF2) algorithms, using 2012 ATLAS $pp$ collision data at $\sqrt{s} = 8$ TeV and 25 ns bunch spacing. The average number of interactions per bunch crossing was 11.3 for this event sample (around 25 millions entries). The COF method is resilient to out-of-time signals and, therefore, leads to a better energy resolution than OF2. Its design is luminosity independent and requires only the information of the pulse shape and pedestal value to compute the 7 amplitudes associated to the 7 samples of the read-out. In this figure only the central sample reconstruction is shown. From: J.M. Seixas (ATLAS), private communication

works best for Gaussian noise. Pile-up has the tendency to add positive or negative tails to the noise distribution, which thus becomes non-Gaussian. The extent of these effects depends on the number of underlying events, and thus leads to a luminosity dependence.

Recently, a method has been proposed that is in principle independent of the luminosity. It is based on a deconvolution process of the same type used in digital processing for communication channel equalization, and aims to fully recover the target signal, rather than estimate its amplitude from pulse sampling [21]. This method has been tested with experimental data obtained with the ATLAS TileCal hadronic calorimeter. Figure 10.5 shows some results from this work.

# References

1. Orito, S., et al.: Nucl. Instrum. Methods **216**, 439 (1983)
2. Akrawy, M.A., et al.: Nucl. Instrum. Methods **A290**, 76 (1990)
3. Checchia, P., et al.: Nucl. Instrum. Methods **A275**, 49 (1989)
4. Bajanov, N.A., et al.: Nucl. Instrum. Methods **A442**, 146 (2000)
5. Bell, K.W., et al.: Nucl. Instrum. Methods **A504**, 255 (2003)

6. Apollinari, G., et al.: Nucl. Instrum. Methods **A311**, 520 (1992)
7. Bertolucci, S., et al.: Nucl. Instrum. Methods **A254**, 561 (1987)
8. Cumalat, J.P., et al.: Nucl. Instrum. Methods **A293**, 606 (1990)
9. Blömker, D., et al.: Nucl. Instrum. Methods **A311**, 505 (1992)
10. Manuisch, J., et al.: Nucl. Instrum. Methods **A312**, 451 (1992)
11. Bertoldi, M., et al.: Nucl. Instrum. Methods **A386**, 301 (1997)
12. Kunori, S.: Proceedings of 7th International Conference on Calorimetry in High Energy Physics, Tucson, Arizona, p. 224. World Scientific, Singapore (1997)
13. CMS Collaboration: Technical proposal for the phase-II upgrade of the Compact Muon Solenoid, CERN-LHCC-2015-10 (2015)
14. ATLAS Collaboration: Letter of Intent for the phase-II upgrade of ATLAS, CERN-LHCC-1-23. CERN, Geneva (2012)
15. Adams, B.W., et al.: (2016). arXiv:1603.01843 [physics.ins-det]
16. Barnyakov, A.Yu., et al.: Nucl. Instrum. Methods **A879**, 6 (2018)
17. White, S.: Nucl. Instrum. Methods **A912**, 298 (2018)
18. Winn, D.R., Onel, Y.: J. Phys. Conf. Ser. **404**, 012021 (2012)
19. Cleland, W.E., Stern, E.G.: Nucl. Instrum. Methods **A338**, 467 (1994)
20. Radeka, V., Rescia, S.: Nucl. Instrum. Methods **A265**, 228 (1988)
21. de Filha, L.M., A., et al.: IEEE Trans. Nucl. Sci. **NS-62**, 3265 (2015)

# Part V
# The State of the Art

# Chapter 11
# Calorimeter Performance

Calorimeters exist in a wide variety and are used in very different types of experiments. The design of a particular calorimeter system is usually driven by requirements stemming from the physics goals of the experiment, and by the available budget. These factors lead to large differences in performance between the various calorimeter systems used in particle physics experiments.

In this chapter we review some of the aspects of the performance of calorimeter systems. The factors that determine and limit various aspects of this performance are described, and results from some representative calorimeters are presented. The possibilities for using calorimeter information for particle identification are also discussed.

## 11.1 Energy Measurement

In Chap. 6, the factors that contribute to and determine the energy resolution of calorimeters are discussed in detail. It is clear what is needed to set new records for calorimetric energy resolution, both for what concerns the detection of electrons/photons and for hadrons/jets. However, in modern collider experiments, design choices are often driven by considerations other than the desire to obtain the best possible energy resolution. Assuming that the performance is adequate for achieving the scientific goals of the experiment, the most important practical considerations when choosing a calorimeter system include:

- The cost,
- The size, which may affect the cost of other components of the detector system, and
- The expected lifetime, in view of radiation and other environmental conditions.

---

The original version of this chapter was revised: Figure 11.5 has been corrected. The correction to this chapter can be found at https://doi.org/10.1007/978-3-030-23653-3_14.

© Springer Nature Switzerland AG 2019

M. Livan and R. Wigmans, *Calorimetry for Collider Physics, an Introduction*,
UNITEXT for Physics, https://doi.org/10.1007/978-3-030-23653-3_11

**Fig. 11.1** The em energy
resolution of the Belle-II
CsI(Tl) calorimeter. From:
Ikeda H. et al. (2000). *Nucl.
Instr. and Meth.* **A441**, 401

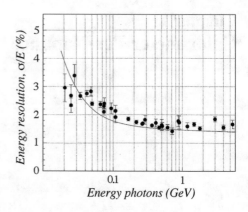

The last point puts systems based on scintillating crystals at a disadvantage, especially at colliders that involve hadrons. These crystals are, apart from very expensive, also very sensitive to ionizing radiation. The CMS experiment, which uses $PbWO_4$ crystals as their em calorimeter experienced this in a big way. Even though these crystals were initially chosen based on their supposed radiation hardness, CMS has to replace a substantial fraction of them after having received only a few percent of the envisaged total integrated luminosity. Crystal calorimeters are still used in some experiments at $e^+e^-$ colliders, such as KEKB, where the event rates are many orders of magnitude smaller than at colliders involving hadron beams. Figure 11.1 shows the energy resolution for photons detected in the Belle-II calorimeter, which consists of CsI(Tl) crystals [1].

A good and increasingly popular alternative is provided by liquified noble gases. Both the ionization charge and the scintillation produced in the absorption process may be used as the source of the signals in this case. Radiation damage is not much of a concern since the liquid can be easily replaced, but the fact that these devices have to operate in cryogenic conditions, as well as the VUV nature of the scintillation ($\sim$100 nm), are the cause of challenging complications.

An example of a homogeneous em calorimeter that detects the *ionization charge* generated by the shower particles can be found in Novosibirsk, where a 70-ton liquid-krypton detector (KEDR) operates at the electron–positron collider VEPP-4M [2]. The energy resolution of this device was measured with positrons on a 400 kg prototype. The measured $\sigma/E$ values ranged from $\sim$5.7% at 0.13 GeV to 1.7% at 1.2 GeV [3].

The NA62 experiment operates a $27X_0$ (1.25 m) deep LKr calorimeter at CERN's SPS, which earlier served the predecessor experiment NA48 [4]. Its em energy resolution was measured with electrons at high energies (10–80 GeV) [5]. The results are shown in Fig. 11.2.

Also LAr and LXe are used in this mode. The CMD-3 Collaboration operates a 400-liter LXe detector at the $e^+e^-$ Collider in Novosibirsk [2]. For a 40-liter LXe device, a resolution was reported of $3.4\%/\sqrt{E}$, for electrons in the energy range 1–6 GeV [6]. Because of the small size of this detector, shower leakage probably

**Fig. 11.2** The em energy resolution of the NA62 liquid-krypton calorimeter. From: Barr, G.D. et al. (1996). *Nucl. Instr. and Meth.* **A370**, 413

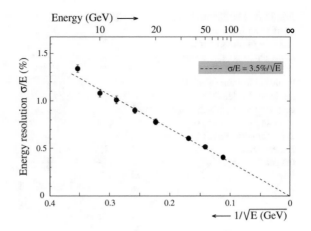

contributed significantly to this result. The latter problem does not play a role for the multi-ton LAr detectors ICARUS, MicroBooNE and DUNE. Despite the very long radiation length (14 cm), these detectors are sufficiently large to fully contain em showers (Fig. 1.8b). A first attempt to measure the energy of em showers developing in liquid argon was performed by members of the ICARUS collaboration, who used signals from such showers in their 600-ton detector to select $\pi^0 \rightarrow \gamma\gamma$ events [7]. Using a restricted sample of "clean" events with an average energy of 700 MeV, they measured the $\pi^0$ mass with a resolution of 16%.

By far the best performance for calorimetric hadron detection is achieved with sampling calorimeters. The reasons for this are spelled out in detail in Chap. 7. The overwhelmingly dominating role of fluctuations in the em shower content (and the related fluctuations in invisible energy) on the hadronic energy resolution of calorimeters is crystal clear. This is perhaps most dramatically illustrated with Fig. 11.3a, which shows the energy resolution for pions in a very large, *fully homogeneous* calorimeter consisting of 60 tons of mineral oil doped with scintillating agents [9]. All other sources of fluctuations, such as sampling fluctuations, have been completely eliminated in this device. The energy resolution turned out to be only weakly dependent on the energy of the pions and did not drop below $\sim$10%, even at the highest energies (150 GeV) at which this detector was tested. For comparison, the resolution of SPACAL, a calorimeter in which only 2.3% of the pion energy was sampled, was about three times better at this energy (Fig. 11.3b). Even better performance was reported by RD52, for their prototype dual-readout fiber calorimeter (Fig. 8.5).

Experimental data make it very clear that there is a price to be paid, in terms of degraded hadronic performance, when the calorimeter system is designed for optimal em energy resolution. This is because of the large *e/h* values typical for good em calorimeters. A case in point is CMS, where the em calorimeter section has an *e/h* value of 2.4. The result of this choice is not only an exceptionally poor hadronic energy resolution, but also very substantial signal non-linearities, and large response differences depending on the starting point of the showers (Fig. 7.6).

**Fig. 11.3** The hadronic
energy resolution as a
function of energy for a
homogeneous calorimeter
consisting of 60 tons of
liquid scintillator (**a**) and for
the SPACAL lead/fiber
calorimeter (**b**), which had a
sampling fraction of only
2.3% for showers [8].
Experimental data from [9]
(**a**) and [10] (**b**)

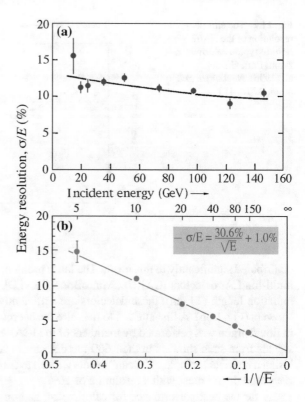

## 11.2   The Other Components of the Four-Vector

### 11.2.1   The Center-of-Gravity of the Showers

The most frequently used method to determine the position of a particle that showers
in a calorimeter is by reconstructing the center of gravity $(\bar{x}, \bar{y})$ of the energies $E_i$
deposited in the various detector cells (with coordinates $x_i, y_i$) that contribute to the
signal:

$$\bar{x} = \frac{\sum_i x_i E_i}{\sum_i E_i} \tag{11.1}$$

and a similar expression for the $\bar{y}$ coordinate.

However, when calculated in this way, the impact points of high-energy electrons
and photons tend to be systematically shifted towards the center of the cell hit by
the showering particle [11–13]. This is a consequence of the steeply falling lateral
em shower profile. Typically, the energy density decreases by an order of magnitude
over a radial distance of only a few cm (see, for example, Fig. 5.5). As a result, in
practical calorimeters, at least half of the shower energy is typically deposited in one

**Fig. 11.4** Signals from 80 GeV electrons, recorded in neighboring calorimeter cells, as a function of the (y coordinate of the) impact point of the particles. From: Acosta, D. et al. (1991). *Nucl. Instr. and Meth.* **A305**, 55

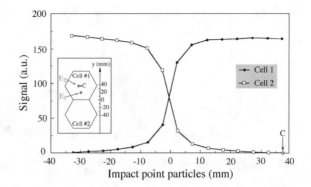

individual calorimeter cell, and the exact percentage is only weakly dependent on the location of the impact point.

This effect is illustrated in Fig. 11.4, which shows the signals generated by 80 GeV showering electrons in neighboring cells of the SPACAL calorimeter [13]. Only in events in which the impact point was located close to the boundary between different calorimeter cells was the shower energy really shared between different cells. In these cases, the particle's position was most precisely and most accurately measured. In other situations, one cell received a lion's share, and the remaining "halo" was shared among the surrounding cells.

Consider an event ($E_1$) in which the showering particle entered the calorimeter halfway between the center of a cell (say cell #1) and the boundary with another cell (cell #2), e.g., at the position $y \approx 20$ mm in Fig. 11.4. For this event, the fraction of the shower energy deposited in cell #1 would not be very different from the fraction recorded for another event ($E_2$), in which the particle entered in the center of cell #1 ($y \approx 40$ mm). In both cases, the contribution of cell #1 to the sum in Eq. 11.1 would be about the same. In both cases, the coordinates of the particle's impact point would be calculated (with Eq. 11.1) under the assumption that *the entire energy* recorded in cell #1 was deposited in its center (point $C$).

Small differences between the signals recorded in the various surrounding cells would in events of the type $E_1$ thus be the only basis for finding an impact point deviating from the center ($C$) of the hit calorimeter cell. However, the reconstructed impact point would always be located too close to $C$, because the bulk of the signal was mis-attributed as originating from this point.

The effects described above are responsible for the peculiar patterns in Fig. 11.5. This figure shows scatter plots for 80 GeV electrons detected with SPACAL, in which the impact point of the particles, reconstructed from the calorimeter data, is plotted versus the "true" impact point, measured with wire chambers installed upstream of the calorimeter [13].

Figure 11.5a, in which this information is given for the $x'$ coordinate (see Fig. 11.6c for the definition of the coordinate system in these hexagonal cells), shows that the impact point was only correctly reconstructed when the particles entered either in the center of a calorimeter cell ($x' = 0$), or near the point where three hexagonal cells

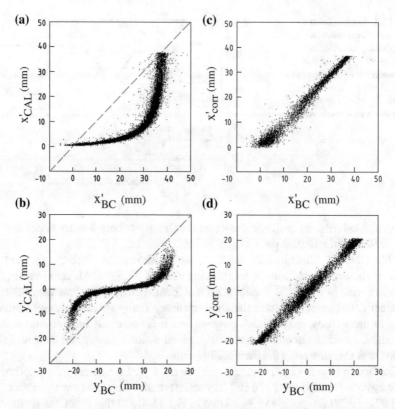

**Fig. 11.5** Scatter plots for 80 GeV electrons detected with the SPACAL calorimeter, showing the relations between the coordinates of the particle's impact point, measured with wire chambers upstream of the calorimeter (horizontal), and determined from the calorimeter data on the basis of the center-of-gravity method (vertical). The data shown in **a** and **c** concern the $x'$ coordinate, **b** and **d** refer to the $y'$ coordinate. The plots in **c** and **d** were obtained after applying the corrections from Eq. 11.2 to the calorimeter data. See text for details. From: Acosta, D. et al. (1991). *Nucl. Instr. and Meth.* **A305**, 55

joined ($x' \approx 40$ mm). In these cases, the points in the scatter plot cluster around the dashed line, which represents the equality of the $x'$ values reconstructed on the basis of the calorimeter data on the one hand and the upstream wire chambers on the other.

However, if the electron entered the calorimeter anywhere else than in the mentioned areas, then the impact point was always reconstructed too close to the cell's center. Figure 11.5a shows that practically all events with (true) impact points in a cylinder with a radius of about 25 mm around the cell's center were reconstructed as entering very close to the calorimeter center ($x'_{CAL} = 0$). These events constitute the horizontal band in this figure. The systematic mismeasurement of the particle position could thus be as large as 30 mm when the calorimeter data were used at face value.

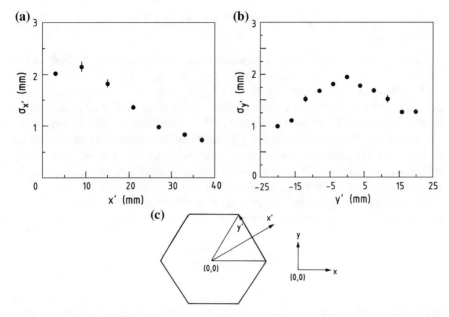

**Fig. 11.6** The position resolution for 80 GeV electrons in SPACAL as a function of $x'$ (**a**) and $y'$ (**b**). The coordinates $(x', y')$ used for the hexagonal SPACAL geometry (**c**), and their relationship to the cartesian $(x, y)$ coordinates. From: Acosta, D. et al. (1991). *Nucl. Instr. and Meth.* **A305**, 55

Figure 11.5b shows similar data for the $y'$ coordinate. Also here, the experimental points cluster in a horizontal band (around the calorimeter center, $y'_{CAL} = 0$), and only near the boundaries with neighboring modules were the impact points correctly reconstructed. In the area between the center and these boundaries, the $y'$ coordinate was also systematically mismeasured when the calorimeter data were used at face value.

This problem can be solved in several ways. In one method that is sometimes applied, a larger weight is given to cells in which only a small fraction of the shower energy is deposited. In another method, the position of the center of gravity found with Eq. 11.1 is shifted, using an empirical algorithm. For example, an algorithm of the type

$$x'_{corr} = A \arctan(Bx') \tag{11.2}$$

applied to the data from Fig. 11.5a, reproduced the $x$ coordinate of the impact points, on average, very well (Fig. 11.5c). A similar algorithm changed the picture for the $y'$ coordinate (Fig. 11.5b) to the pattern shown in Fig. 11.5d.

The position resolution is given by the widths of the bands in Fig. 11.5c (for the $x'$ coordinate) and d ($y'$). Not surprisingly, this width was found to depend on the impact point of the particles. The smallest resolutions were obtained in the boundary areas between different calorimeter cells (Fig. 11.6). Averaged over one cell, the position resolutions $\sigma_{x'}$ and $\sigma_{y'}$ were found to be 1.8 mm and 1.6 mm, respectively, for 80 GeV electrons.

The position resolution may be expected to scale with $1/\sqrt{E}$ on the basis of the following arguments. The energy deposit $E_i$ in each cell $i$ has a relative precision $\sigma_i/E_i$. This relative precision improves as $1/\sqrt{E}$ with the total shower energy $E$, *provided that the (average) shower profile stays the same*. In that case, the energy sharing between the various calorimeter cells $i$ is, on average, independent of the shower energy. If all the terms in Eq. 11.1 have a relative precision that scales with $1/\sqrt{E}$, and if the relative contributions of the individual terms to the sum are energy independent, then the relative precision of the final result (i.e., the sum of all the terms, or the value of the position coordinate) must also scale with $1/\sqrt{E}$.

This is indeed in agreement with experimental observations. For electrons entering the detector in the center of a cell, SPACAL measured a position resolution

$$\sigma_{y'} = \frac{17.1 \text{ mm}}{\sqrt{E \text{ (GeV)}}} \tag{11.3}$$

and a similar result was obtained for the $x'$ coordinate. The cell center represents the worst possible case for what concerns the position resolution. Averaged over the entire surface, the position resolution was found to be about 20% better than indicated in Eq. 11.3 [13].

The position resolution is not only determined by the energy resolution, but also by the cell size. The smaller the cell size (measured in the relevant units of the Molière radius, $\rho_M$), the more cells contribute to the signals, and the more accurately the shower's center of gravity can be determined.

The cell size of SPACAL, which had an effective radius of $1.9\rho_M$, was by no means optimized for electron impact-point determination, since an electron hitting a cell in its central region deposited typically $\sim$95% of its shower energy in this one cell. The position resolution of the RD1 projective prototype fiber calorimeter, which had a cell size with an effective radius of $1.1\rho_M$, and an em energy resolution similar to SPACAL, was found to be smaller by more than a factor of two: $\sigma_{x,y} = 7.5$ mm/$\sqrt{E}$ [14].

## 11.2.2  Localization Through Timing

A completely different way of determining the position of the showering particles was used in the KLOE experiment at the $\phi$ factory DAPHNE (Frascati, Italy), which studied *CP* violation in $K^0$ decays and several other physics topics [15]. When the $\phi$s, produced at rest in $e^+e^-$ collisions at $\sqrt{s} = 1.020$ GeV, decay into two $K^0$s these carry a momentum of 110 MeV/c each.

This experiment operated a $4\pi$ electromagnetic calorimeter [16, 17] built to perform three major tasks:

- Determine the vertex for the decays $K^0 \to \pi^0 \pi^0$.
- Reject the $K^0 \to \pi^0 \pi^0 \pi^0$ events with good efficiency.
- Provide the trigger for the experiment.

In order to perform these tasks, the calorimeter needed to measure the $\gamma$s from the decays $\pi^0 \to \gamma\gamma$ with good precision. These $\gamma$s typically had energies in the range from 50 to 300 MeV.

The calorimeter consisted of scintillating fibers embedded in lead. The barrel section of the calorimeter had a length of 3.75 m and an outer diameter of 4 m, with the fibers running parallel to the beam line. Because of the very fast signals generated by the fibers, the time resolution was extremely good: 34 ps/$\sqrt{E}$ [16].

Light travels through these fibers at a speed of $\sim$17 cm ns$^{-1}$. By comparing the arrival times of the scintillation light at both ends of the fibers, the shower position along the fiber direction could thus be determined with a precision of only 1–3 cm, for the $\gamma$s of interest (Fig. 11.7, righthand scale).

In this experiment, a good localization of the $\gamma$s was very important to identify the parent particle. The invariant mass of a particle decaying into two $\gamma$s is given by

$$M = \sqrt{2E_1 E_2 (1 - \cos\theta_{12})} \tag{11.4}$$

The precision with which the mass can be measured is thus not only determined by the energy resolution, i.e., the measurement uncertainty on the $\gamma$ energies $E_1$ and $E_2$, but also by the relative uncertainty on the angle ($\theta_{12}$) between the directions of these $\gamma$s. The latter uncertainly is of course directly affected by the precision of the mentioned localization procedure.

Figure 9.10 shows the invariant mass distributions of $\gamma\gamma$ and $\gamma\gamma\gamma\gamma$ combinations recorded in the $e^+ e^-$ collisions at the $\phi$ resonance in DAPHNE. The masses of the

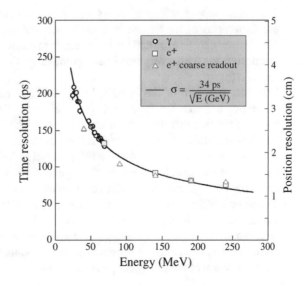

**Fig. 11.7** Time resolution (left-hand scale) of the KLOE calorimeter for signals generated by $\gamma$s as a function of energy, and the position resolution (right-hand scale) in the direction along the fiber that can be achieved by comparing the arrival times of the shower light at both ends of the fiber. From: Antonelli, A. et al. (1995). *Nucl. Instr. and Meth.* **A354**, 352

$\pi^0$, $\eta$ and $K_S^0$, which decay in these exclusive modes, were reconstructed, on average, to within 1% of their established values, with good resolutions [17].

A good localization of the $\gamma$s is thus very important to identify the parent particle. While CMS emphasized excellent energy resolution for em showers in its design of the experiment, at the expense of degraded hadronic performance, ATLAS concentrated its efforts also on the localization issue. As a result, the mass resolution for the Higgs bosons, measured in the $H^0 \rightarrow \gamma\gamma$ decay channel, turned out to be very similar in both experiments.

## 11.3   Particle Identification

### 11.3.1   Electron Identification in Practice

Identification of electrons is very important for many physics studies. Owing to differences in the characteristic energy deposit profiles, and to other features of shower development, calorimeter signals may be used to identify the particles that caused the signals. The methods used for this purpose are based on:

1. *The longitudinal shower information.* Especially in high-$Z$ materials, the difference between the values of the scaling variables that define the longitudinal development of em ($X_0$) and hadronic ($\lambda_{int}$) showers is very large (Fig. 3.8). Such materials are therefore ideally suited for making very efficient preshower detectors, as illustrated in Fig. 5.15. Electron identification is one of the most important reasons why almost all calorimeters in modern experiments consist of separate em and hadronic sections, with the em section typically made of high-$Z$ absorber material.
2. *The lateral shower information.* As was shown in Chap. 5, the lateral profiles of em showers are characterized by an extremely collimated central core (Fig. 5.6). This feature may be explored to obtain excellent electron identification, especially in calorimeters with a very fine lateral granularity. This is illustrated in Figs. 12.9 and 12.10.
3. *The time structure of the signals.* In Fig. 8.13, an example was shown of the difference between the time structure of electron and pion signals that was a result of the fact that the light in the active material (optical fibers in this case) traveled at a lower speed than the shower particles that created this light. The difference in longitudinal shower development thus led to a difference in the time structure of the signals. However, the signals from pions may also have a time structure that is *intrinsically* different from electron ones. This is a consequence of the contribution of evaporation neutrons, which release their kinetic energy on a time scale of $\sim$10 ns, as illustrated in Fig. 11.8.
4. *A comparison between the energy and the momentum of the particle.* This is known as the $E/p$ method. The energy, measured calorimetrically, is compared to the momentum of the (charged) particle, measured from the curvature of its track

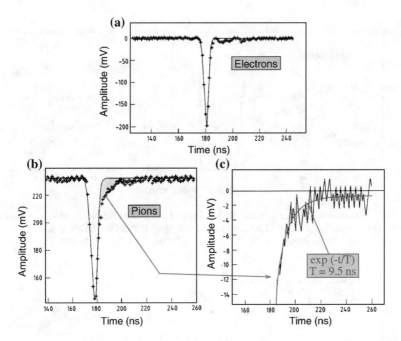

**Fig. 11.8** Typical calorimeter signals for 150 GeV electrons (**a**) and pions (**b**) measured with the SPACAL calorimeter. The pion signal exhibits a clear exponential tail with a time constant of ~10 ns (**c**). The $t = 0$ point is arbitrary and the bin size is 1 ns. From: Acosta, D. et al. (1991). *Nucl. Instr. and Meth.* **A302**, 36

in a magnetic field upstream of the calorimeter. For electrons depositing their entire energy in (the em section of) the calorimeter, this ratio is approximately 1.0. Pions deposit typically only a fraction of their energy in the em calorimeter section and, therefore, yield an *E/p* value less than 1.0.

An example of results obtained with this method is given in Fig. 11.9 [18]. Electrons that deposited their entire energy in this calorimeter showed up as a clear peak centered around 1.0 in these distributions. Events in which significant leakage occurred, i.e., electrons that entered the calorimeter near its boundaries, or pion events that were not removed by various cuts that were applied, were characterized by *E/p* values less than 1.0. The width of the peak, and thus the rejection power of this method, is determined both by the energy resolution of the calorimeter and by the momentum resolution of the spectrometer. The energy resolution improves with energy, while the momentum resolution degrades ($\Delta p/p$ is proportional to $p$). Therefore, the latter dominated the rejection power of this method at high energies. This is illustrated by the fact that the peak in Fig. 11.9b (for the momentum bin 65–70 GeV/c) is broader than that in Fig. 11.9a (55–60 GeV/c).

In practice, often a combination of these methods is used. In addition to a magnetic spectrometer, information from other detectors, based on the detection of Čerenkov or transition radiation, may be helpful too.

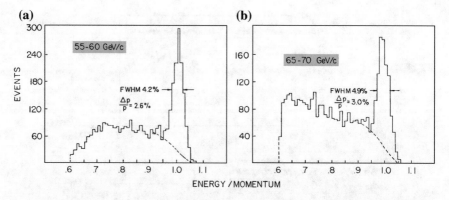

**Fig. 11.9** Distribution of the *E/p* values of a sample of candidate electrons recorded in the E-70 experiment at Fermilab, for particles with momenta ranging from 55 to 60 GeV/c (**a**) and from 65 to 70 GeV/c (**b**). From: Appel, J.A. et al. (1975). *Nucl. Instr. and Meth.* **127**, 495

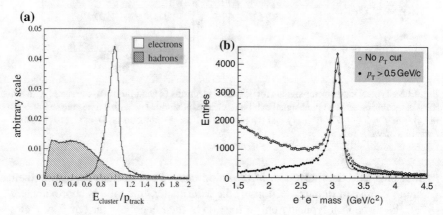

**Fig. 11.10** The ratio of the energy measured by the ECAL to the momentum measured by the magnetic spectrometer in the LHCb experiment (**a**). Invariant mass plots for the $e^+e^-$ pairs in a sample of $B_s^0 \rightarrow J/\psi\phi$ signal events, showing the effect of a cut $p_T > 0.5$ GeV/c for the $e^\pm$ candidates (**b**) on the reconstructed $J/\psi$ peak. From: Alves, A.A. et al. (2008). *JINST* **3**, S08005

An example of a modern experiment that uses such techniques is LHCb, which studies *CP* violation and rare decays of particles containing *b* quarks at the LHC [19]. Identification of electrons is extremely important in these studies. Apart from a magnetic spectrometer, this experiment also uses a system of powerful Ring Imaging Cherenkov counters to identify electrons buried in the huge background of heavier particles [20]. Figure 11.10a shows the *E/p* signal ratio for electrons and hadrons. The low-energy tail in the electron peak is the result of bremsstrahlung losses in the material upstream of the calorimeter. The importance of correctly identifying electrons (and positrons) in this experiment is illustrated in Fig. 11.10b, which shows the invariant mass distribution of $e^+e^-$ pairs in a sample of $B_s^0 \rightarrow J/\psi\phi$ events, with the $J/\psi$ decaying into an electron–positron pair. The background in this event sample is mainly due to pion tracks with low transverse momentum, and could be efficiently removed with a $p_T$ cut that barely affected the electrons. It turns out that the average efficiency for identifying electrons from $J/\psi \rightarrow e^+e^-$ decays in such events is 95%, with a pion mis-identification fraction of 0.7%.

## 11.3.2 Longitudinally Unsegmented Calorimeters

The RD52 fiber calorimeter is longitudinally unsegmented, it does not consist of separate electromagnetic and hadronic sections. It is calibrated with electrons, and the calibration constants established in this way also provide the correct energy for hadronic showers developing in it. This eliminates one of the main disadvantages of longitudinal segmentation, i.e., the problems associated with the intercalibration of the signals from different longitudinal sections (Chap. 9). Another advantage derives from absence of the necessity to transport signals from the upstream part of the calorimeter to the outside world. This allows for a much more homogeneous and hermetic detector structure in a $4\pi$ experiment, with fewer "dead areas."

Despite the absence of longitudinal segmentation, the signals provided by the RD52 fiber calorimeter offer several excellent possibilities to distinguish between different types of particles, and especially between electrons and hadrons. Identification of isolated electrons, pions and muons would be of particular importance for the study of the decay of Higgs bosons into pairs of $\tau$ leptons, if a calorimeter of this type were to be used in an experiment at a future Higgs factory.

Figure 11.11 illustrates the effects of the different identification methods [21]:

1. There are large differences in *lateral* shower size, which can be used to distinguish between em and hadron showers. One advantage of the RD52 calorimeter structure is that the lateral granularity can be made arbitrarily small, one can make the tower size (defined by the number of fibers connected to one readout element) as large or small as desired. Figure 11.11a shows the distributions of the fraction of the shower energy deposited in a RD52 tower located on the shower axis, for 60 GeV electrons and pions.
2. The availability of both scintillation and Čerenkov signals from the same events offers opportunities to distinguish between em and hadronic showers. For example, the ratio between the two signals is 1.0 for electrons (which are used to calibrate the signals!) and smaller than 1.0 for hadrons, to an extent determined by $f_{em}$ and $e/h$. Figure 11.11b shows the distributions of the Čerenkov/scintillation signal ratio for 60 GeV electrons and pions.
3. The next two methods are based on the fact that the light produced in the fibers travels at a lower speed $(c/n)$ than the particles responsible for the production of that light, which typically travel at $c$. As a result, the deeper inside the calorimeter the light is produced, the earlier it arrives at the PMT. Since the light from hadron showers is typically produced much deeper inside the calorimeter, the PMT signals start earlier than for em showers, which all produce light close to the front

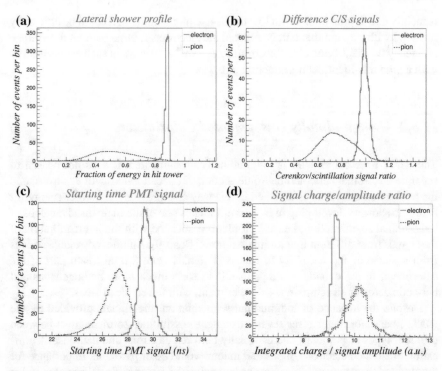

**Fig. 11.11** Effects of four different shower characteristics that may be used to distinguish between electron and hadron showers in the longitudinally unsegmented RD52 lead-fiber calorimeter. Shown are the fraction of the total signal recorded by the tower in which the particle entered (**a**), the ratio of the Čerenkov and scintillation signals of the event (**b**), the starting time of the signal in the PMT, measured with respect to an upstream trigger signal (**c**), and the ratio of the total integrated charge and the amplitude of the signal (**d**). Data obtained with 60 GeV particle beams. From: Akchurin, N. et al. (2014). *Nucl. Instr. and Meth.* **A735**, 120

face of the calorimeter. Figure 11.11c shows distributions of the starting time of the PMT signals for 60 GeV electron and pion showers.

4. The same phenomenon also leads to a larger width of the hadron signals, since the light is produced over a much larger region in depth than for electrons. Therefore, the ratio of the integrated charge and the signal amplitude is typically larger for hadron showers. Figure 11.11d shows distributions of that ratio for showers induced by 60 GeV electrons and pions.[1]

One may wonder to what extent the different methods mentioned above are correlated, in other words to what extent the mis-identified particles are either the same or different ones for each method. It turned out that by combining different $e/\pi$ sep-

---

[1]This result may be compared with Fig. 8.14, which shows the difference between the widths of electron and pion signals in the SPACAL calorimeter. However, in that case, the differences were greatly enhanced by the aluminized upstream ends of the fibers. This had a much larger effect on the signal structure of the hadron showers than for the electron ones.

aration methods, important improvements could be achieved in the capability of the longitudinally unsegmented calorimeter to identify electrons with minimal contamination of mis-identified particles. A multivariate neural network analysis showed that the best $e/\pi$ separation achievable with the variables used for the 60 GeV beams was 99.8% electron identification with 0.2% pion misidentification. Further improvements may be expected by including the full time structure information of the pulses, especially if the upstream ends of the fibers are made reflective [22].

The longitudinally unsegmented RD52 fiber calorimeter can thus be used to identify electrons with a very high degree of accuracy. Elimination of longitudinal segmentation offers the possibility to make a finer lateral segmentation with the same number of electronic readout channels. This has many potential benefits. A fine lateral segmentation is crucial for recognizing closely spaced particles as separate entities. Because of the extremely collimated nature of em showers (Fig. 5.6), it is also a crucial tool for recognizing electrons in the vicinity of other showering particles, as well as for the identification of electrons in general. Unlike the vast majority of other calorimeter structures used in practice, the RD52 fiber calorimeter offers almost limitless possibilities for lateral segmentation. If so desired, one could read out every individual fiber separately. Modern silicon PM technology certainly makes that a realistic possibility, as illustrated by Figs. 12.9 and 12.10.

## 11.4 Tricks to Obtain Useful Details of the Shower Development

### 11.4.1 Exploiting the Wonderful Features of Čerenkov Light

Čerenkov light has a number of specific properties that offer unique possibilities for application in calorimetry:

1. It is directional.
2. It is instantaneous.
3. Its spectrum is quite different from that of typical scintillators.

One unexpected consequence of the directionality of the Čerenkov light concerned the detection of muons with the dual-readout fiber calorimeter discussed in Chap. 8. Simultaneous detection of the scintillation and Čerenkov light produced in this instrument turned out to have unique beneficial aspects for the detection of these particles. Figure 11.12 shows the average signals from muons traversing the DREAM calorimeter along the fiber direction [23]. The gradual increase of the response with the muon energy is a result of the increased contribution of radiative energy loss (bremsstrahlung) to the signals. The Čerenkov fibers are *only* sensitive to this energy loss component, since the primary Čerenkov radiation emitted by the muons falls outside the numerical aperture of the fibers. The constant (energy-independent) difference between the total signals observed in the scintillating and Čerenkov fibers

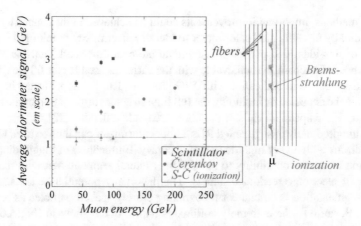

**Fig. 11.12** Average values of the scintillation and Čerenkov signals from muons traversing the DREAM calorimeter, as a function of the muon energy. Also shown is the difference between these signals. All values are expressed in units of GeV, as determined by the electron calibration of the calorimeter. From: Akchurin, N. et al. (2004). *Nucl. Instr. and Meth.* **A533**, 305

thus represents the non-radiative component of the muon's energy loss. Since the signals from both types of fibers were calibrated with em showers, their responses to the radiative component were equal. This is a unique example of a detector that separates the energy loss by muons into radiative and non-radiative components.

The difference in the time structure of the signals is an important characteristic that can be used to distinguish between the scintillation and Čerenkov components of the light produced by high-energy particles in crystals. And of course, the larger the difference in the time structure, the better the separation works. The RD52 collaboration managed to improve the applicability of $PbWO_4$ crystals for dual-readout calorimetry by doping them with small amounts, $\mathcal{O}(1\%)$, of molybdenum [24]. This had two beneficial effects: it increased the decay time of the scintillation light and it shifted the spectrum of the emitted scintillation light to larger wavelengths.

The effects of that are illustrated in Fig. 11.13, which shows the calorimeter signals generated by 50 GeV electrons traversing a crystal of this type. This crystal was oriented such as to maximize the relative fraction of Čerenkov light in the detected signals. By selecting the UV light by means of an optical filter, almost the entire detected signal was due to (prompt) Čerenkov light, while a yellow transmission filter predominantly selected scintillation light, which had a decay time of ~26 ns as a result of the Mo-doping. Whereas the differences in angular dependence were very suitable to demonstrate that some of the light generated in these crystals is actually the result of the Čerenkov mechanism, the combination of time structure and spectral characteristics provides powerful tools to separate the two types of light in real time. One does not even have to equip the calorimeter with two different light detectors for that. This was demonstrated with a calorimeter consisting of bismuth germanate ($Bi_4Ge_3O_{12}$, or BGO) crystals [25].

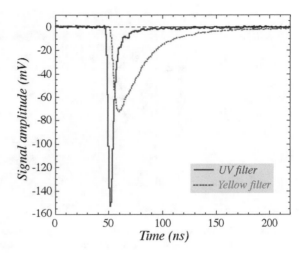

**Fig. 11.13** Average time structure of the signals from a PbWO$_4$ crystal doped with 1% Mo, generated by 50 GeV electrons. The angle $\theta$ was 30° in these measurements. Shown are the results obtained with UV and yellow filters, respectively. From: Akchurin, N. et al. (2009). *Nucl. Instr. and Meth.* **A604**, 512

Even though Čerenkov radiation represents a very tiny fraction of the light produced by these crystals, it is relatively easy to separate and extract it from the signals. The much longer scintillation decay time (300 ns) and the spectral difference are responsible for that.[2] Figure 11.14 shows the time structures of signals from a BGO calorimeter recorded with a UV filter. The "prompt" component observed in the ultraviolet signal is due to Čerenkov light. A small fraction of the scintillation light also passes through the UV filter. This offers the possibility to obtain all needed information from only one signal. An external trigger opens two gates: one narrow (10 ns) gate covers the prompt component, the second gate (delayed by 30 ns and 50 ns wide) only contains scintillation light. The latter signal can also be used to determine the contribution of scintillation to the light collected in the narrow gate. In this way, the Čerenkov/scintillation ratio can be measured event-by-event on the basis of one signal only [25].

Figure 11.15 illustrates how this Čerenkov light yield can be measured in practice [26]. It concerns measurements on a PbWO$_4$ crystal doped with 0.3% of molybdenum. This crystal was placed at an angle $\theta = 30°$ with the beam line. One PMT ($R$) was equipped with a UV filter, in order to select the Čerenkov light, for which the detection efficiency is largest at this angle. At the other side of the crystal only scintillation light was detected. EGS4 calculations indicated that the beam particles (50 GeV electrons) deposited on average 0.578 GeV in this crystal, which was slightly thicker than $2X_0$ in this geometry. This made it possible to calibrate the scintillation signals, the distribution of which is shown in Fig. 11.15a. This distribution was subdivided into 20 bins. For each bin, the signal distribution on the opposite side of the crystal, i.e., the Čerenkov side, was measured. The fractional width of this distribution, $\sigma_{rms}/C_{mean}$ , is plotted in Fig. 11.15b versus the average scintillator signal in this bin, or rather versus the inverse square root of this signal ($S^{-1/2}$). It turned out that

---

[2]The BGO scintillation spectrum peaks at 480 nm, while Čerenkov light exhibits a $\lambda^{-2}$ spectrum.

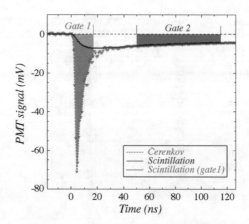

**Fig. 11.14** The time structure of a typical shower signal measured in the BGO em calorimeter equipped with a UV filter. These signals were measured with a sampling oscilloscope, which took a sample every 0.8 ns. The UV BGO signals were used to measure the relative contributions of scintillation light (gate 2) and Čerenkov light (gate 1). From: Akchurin, N. et al. (2009). *Nucl. Instr. and Meth.* **A610**, 488

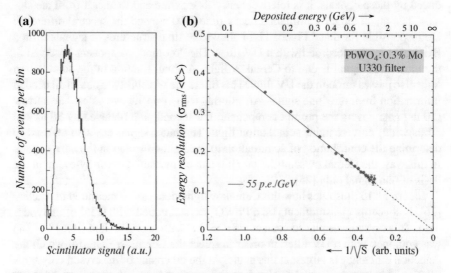

**Fig. 11.15** The scintillation signal distribution for 50 GeV electrons traversing a PbWO$_4$ crystal at $\theta = 30°$ (**a**) and the fractional width of the Čerenkov signal distribution as a function of the amount of energy deposited in the crystal, as derived from the scintillation signal (**b**). The crystal was doped with 0.3% Mo. See the text for more details. From: Akchurin, N. et al. (2010). *Nucl. Instr. and Meth.* **A621**, 212

this fractional width scaled perfectly with this variable, i.e., with $E^{-1/2}$. Since the relationship between the energy $E$ and the scintillation signal $S$ is given by the calibration described above, it was also possible to indicate the energy scale in Fig. 11.15b. This is done on the top horizontal axis. The observed scaling of $\sigma_{rms}/C_{mean}$ with $E^{-1/2}$ means that the energy resolution is completely determined by stochastic processes that obey Poisson statistics. In this case, fluctuations in the Čerenkov light yield were the only stochastic processes that played a role, and therefore the average light yield could be directly determined from this result: 55 photoelectrons per GeV deposited energy. For an energy deposit of 1 GeV, this led to a fractional width of 13.5%, and therefore the contribution of Čerenkov photoelectron statistics amounts to $13.5\%/\sqrt{E}$. This is not much better than what could be achieved in a dedicated fiber sampling calorimeter.

## 11.4.2   A Caveat

The directionality of Čerenkov light is also the reason for an important caveat. As long as the shower particles are isotropically distributed with respect to the direction of incidence of the showering particle, this has no consequences. However, if (some fraction of) the charged particles produced in the absorption process are non-isotropically distributed, the calorimeter signal may depend on this angle of incidence. An extreme consequence of this phenomenon was shown in Fig. 11.12. Muons traversing the dual-readout DREAM fiber calorimeter only produced Čerenkov signals resulting from the radiative shower losses, since light emitted by the muon itself was not trapped within the numerical aperture of the fibers.

This phenomenon may also affect the response of a Čerenkov calorimeter to electromagnetic showers. Figure 11.16a shows the angular distribution of the shower particles through which the energy of a high energy electron is deposited in a (lead) absorber [27]. This distribution contains a sizeable component of more or less isotropically distributed relativistic electrons, i.e., electrons capable of emitting Čerenkov light. These electrons are predominantly produced in Compton scattering. They have "forgotten" the direction of the high-energy particle that initiated the shower and are more or less randomly oriented with respect to that direction. This component leads to a Čerenkov signal in fibers oriented at 0°, or any other angle with respect to the flight path of the showering particles. When the fibers are oriented at the Čerenkov angle ($\theta_C = \arccos(n^{-1}) \approx 46°$) with respect to that flight path, the signals from this type of calorimeter contain, *in addition*, contributions from most of the electrons and positrons produced in the early phase of the shower development, which is dominated by $\gamma \rightarrow e^+e^-$ processes.

Figure 11.16b shows the em response as a function of the angle of incidence of the showering particles for a quartz-fiber calorimeter [28]. This response does show an angular dependence. As expected, the highest response is indeed obtained when the angle of incidence corresponds to the Čerenkov angle. At that angle, the fibers are sensitive to particles that travel in the same direction as the incoming particle. If all

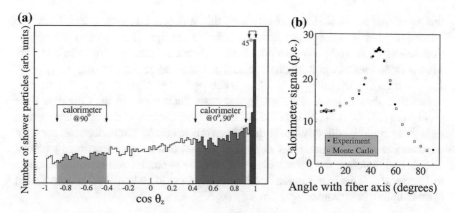

**Fig. 11.16** Angular distribution of the shower particles (electrons and positrons) through which the energy of a 1 GeV electron is absorbed in a lead-based calorimeter. Results of EGS4 Monte Carlo simulations. The angular regions contributing to the signals from calorimeters with quartz fibers oriented at 0°, 45° and 90° are shaded and indicated by arrows (**a**). From: Acosta, D. et al. (1990). *Nucl. Instr. and Meth.* **A294**, 193. The electromagnetic calorimeter response of a fiber calorimeters based on the detection of Čerenkov light, as a function of the angle of incidence of the particles, measured with respect to the fiber axis (**b**). From: Ganel, O. and Wigmans, R. (1995). *Nucl. Instr. and Meth.* **A365**, 104

shower particles that emit Čerenkov light traveled in that direction, then the response of this calorimeter would be zero except for angles of incidence of $46° \pm \Delta\theta$, where $\Delta\theta$ is determined by the numerical aperture of the fibers and typically has values in the 10°–20° range [29].

However, a large fraction of the shower particles through which high-energy electrons and photons deposit their energy in an absorbing structure do not travel in that direction (see Fig. 11.16a). The Čerenkov light emitted by such electrons thus travels at different angles with the fiber axes as well, and the angular dependence of the calorimeter response provides information about the angular distribution of the shower particles with energies above the Čerenkov threshold.

For example, the signal measured at 0°, the orientation needed for calorimeters in colliding-beam experiments, is caused by shower particles traveling at angles of about 45° with the fiber axes, in the forward direction. The response at this angle is smaller than the maximum one, by about a factor of two. Even at an angle of 90°, where the signal from this calorimeter (partially) depends on particles traveling at an angle of about 135° with the fiber axis, i.e., at 45° in the *backward* direction, there is still a significant response. The different angular regions that contribute to the signals from a Čerenkov fiber calorimeter, depending on its orientation, are indicated in Fig. 11.16a. The experimental response curve (Fig. 11.16b) is in excellent agreement with the one derived from Monte Carlo simulations, which follows from the angular distribution of the shower particles [28].

The message from all this is that the calibration of the signals from em showers detected in a Čerenkov fiber calorimeter is, strictly speaking, only valid for particles that enter the calorimeter at angles smaller than $\sim 20°$ with the fiber direction. For particles entering at larger angles, a correction factor will be needed. However, a word of caution is necessary concerning results such as those shown in Fig. 11.16b. The angular dependence of the em response may in reality be less strong than indicated in this figure. That is because measurements performed at angles larger than $20°$ typically concern only the first part of the developing shower. It has been shown that the relative contributions of Compton and photoelectrons to the signals are much larger beyond the shower maximum than in the early part [30]. Therefore, the asymmetry between the response at $0°$ and at $90°$ is probably considerably smaller than it appears to be in Fig. 11.16b. Nevertheless, the arguments listed above for the angular dependence of the Čerenkov response to em showers are perfectly valid.

# References

1. Ikeda, H., et al.: Nucl. Instrum. Methods **A441**, 401 (2000)
2. Peleganchuk, S.: Nucl. Instrum. Methods **A598**, 248 (2009)
3. Aulchenko, V.M., et al.: Nucl. Instrum. Methods **A289**, 468 (1990)
4. Fanti, V., et al. : Nucl. Instrum. Methods **A574**, 433 (2007)
5. Barr, G.D., et al.: Nucl. Instrum. Methods **A370**, 413 (1996)
6. Baranov, A., et al.: Nucl. Instrum. Methods **A294**, 439 (1990)
7. Ankowski, A., et al.: Acta Phys. Pol. **B41**, 103 (2010)
8. Wigmans, R.: Calorimetry—Energy Measurement in Particle Physics, 2nd edn. International Series of Monographs on Physics, vol. 168. Oxford University Press, Oxford (2017)
9. Benvenuti, A., et al.: Nucl. Instrum. Methods **125**, 447 (1975)
10. Acosta, D., et al.: Nucl. Instrum. Methods **A308**, 481 (1991)
11. Akopdjanov, G.A., et al.: Nucl. Instrum. Methods **140**, 441 (1977)
12. Carrington, R.L., et al.: Nucl. Instrum. Methods **163**, 203 (1979)
13. Acosta, D., et al.: Nucl. Instrum. Methods **A305**, 55 (1991)
14. Badier, J., et al.: Nucl. Instrum. Methods **A337**, 326 (1994)
15. Franzini, P., Moulson, M.: Annu. Rev. Nucl. Part. Sci. **56**, 207 (2006)
16. Antonelli, A., et al.: Nucl. Instrum. Methods **A354**, 352 (1995)
17. Adinolfi, M., et al.: Nucl. Instrum. Methods **A482**, 364 (2002)
18. Appel, J.A., et al.: Nucl. Instrum. Methods **127**, 495 (1975)
19. Alves, A.A., et al.: JINST **3**, S08005 (2008)
20. Ypsilantis, T., Séguinot, J.: Nucl. Instrum. Methods **A368**, 229 (1995)
21. Akchurin, N., et al. : Nucl. Instrum. Methods **A735**, 120 (2014)
22. Acosta, D., et al.: Nucl. Instrum. Methods **A302**, 36 (1991)
23. Akchurin, N., et al.: Nucl. Instrum. Methods **A533**, 305 (2004)
24. Akchurin, N., et al.: Nucl. Instrum. Methods **A604**, 512 (2009)
25. Akchurin, N., et al.: Nucl. Instrum. Methods **A610**, 488 (2009)
26. Akchurin, N., et al.: Nucl. Instrum. Methods **A621**, 212 (2010)
27. Acosta, D., et al.: Nucl. Instrum. Methods **A294**, 193 (1990)
28. Ganel, O., Wigmans, R.: Nucl. Instrum. Methods **A365**, 104 (1995)
29. Akchurin, N., Wigmans, R.: Rev. Sci. Instrum. **74**, 2955 (2003)
30. Cardini, A., et al.: Nucl. Instrum. Methods **A808**, 41 (2016)

# Chapter 12
# Particle Flow Analysis

## 12.1 Introduction

A completely different method that has been proposed to meet the challenging requirements of recognizing, reconstructing and separating hadronically decaying *W* and *Z* bosons in future particle physics experiments is the so-called *Particle Flow Analysis* (PFA). This method is based on the combined use of a precision tracker and a highly-granular calorimeter. The idea is that the charged jet particles can be precisely measured with the tracker, while the energy of the neutral particles is measured with the calorimeter. Such methods have indeed successfully been used to improve the mass resolution of hadronically decaying $Z^0$s at LEP [1], and the jet energy resolution using $\gamma$-jet $p_T$ balancing events at CDF [2] and at CMS [3]. Two detector concepts studied in the context of the experimental program for the proposed International Linear Collider (ILC) are based on this method as well [4].

The problem that limits the success of this method is of course that the calorimeter does not know or care whether the particles it absorbs are electrically charged. Therefore, one will have to correct the detected calorimeter signals for the contributions of the charged jet particles. Proponents of this method have advocated a fine granularity as the key to the solution of this "double-counting" problem [5]. However, it has been argued by others that this, for practical geometries, is an illusion [6]. Especially in jets with leading charged particles, the overlap between the showers from individual jet particles makes the fine granularity largely irrelevant.

In order to increase the spatial separation between showers induced by the various jet particles, and thus alleviate the double-counting problem, the concept detectors for the ILC that are based on the PFA principle count on strong solenoidal magnetic fields (4–5 T). Such fields may indeed improve the validity of PFA algorithms, especially at large distances from the vertex, since they open up a collimated beam of particles. It is important to be quantitative in these matters. After having traveled a typical distance of one meter in a 4 T magnetic field, the trajectory of a 10 GeV pion deviates by 6 cm from that of a straight line, i.e., less than one third of a nuclear interaction length (the characteristic length scale for lateral hadronic shower development) in typical

© Springer Nature Switzerland AG 2019

M. Livan and R. Wigmans, *Calorimetry for Collider Physics, an Introduction*,
UNITEXT for Physics, https://doi.org/10.1007/978-3-030-23653-3_12

calorimeters. The field is not always beneficial, since it may also have the effect of bending jet particles with a relatively large transverse momentum with respect to the jet axis *into* the jet core.

Of course, in the absence of reliable Monte Carlo simulations[1] the only way to prove or disprove the advocated merits of the proposed PFA methods is by means of dedicated experiments in realistic prototype studies.

## 12.2   The Importance of Calorimetry for PFA

The first statement in any talk about PFA mentions that 2/3 of the final-state particles constituting a jet are electrically charged, and that the momenta of these particles can be measured extremely precisely. This is of course true, but the implication that the calorimeters of PFA based detector systems don't have to be very good, since they only have to measure one third of the jet energy, is incorrect. In the absence of calorimeter information, based on tracker information alone, the jet energy resolution would be determined by the *fluctuations* in the fraction of the total jet energy that is carried by the charged fragments. This issue was studied by Lobban et al. [6], who found that these event-to-event fluctuations are very large. Depending on the jet fragmentation algorithm, the $\sigma_{rms}$ of the energy fraction carried by charged particles was found to be 25–30% of the average value, *independent of the jet energy*.

One may wonder why these fluctuations do not become smaller at higher energies, given the fact that the *number* of jet fragments increases. The reason for this is that the observed increase in multiplicity is uniquely caused by the addition of more *soft* particles. The bulk of the jet energy is invariably carried by a small number of the most energetic particles. This means that the fraction of the jet energy carried by charged particles is strongly dependent on the extent to which these particles participate in the "leading" component of the jet. Therefore, the event-to-event fluctuations in this fraction are large and do *not* become significantly smaller as the jet energy increases.

As an aside, we mention that the same argument thus necessarily also applies for the event-to-event fluctuations in the fraction of the neutral particles (mainly $\pi^0$s). These fluctuations are responsible for the poor jet energy resolution of non-compensating calorimeters, especially at high energy, since the response of such calorimeters is usually considerably larger for em showers than for non-em ones.

In the absence of a calorimeter, one should therefore not expect to be able to measure jet energy resolutions better than 25–30% on the basis of tracker information alone, *at any energy*. And since the contributions of showering charged particles to the calorimeter signals have to be discounted properly for the PFA method to work, the quality of the calorimeter information is in practice very important, if one aims to achieve the performance needed to separate hadronically decaying $W$ and $Z$ bosons.

---

[1]Concern about the absence of reliable simulations for hadronic shower development was the main reason for a special workshop held at Fermilab in 2006 [7]. To our knowledge, the fundamental problems addressed at this workshop, e.g., with regard to the hadronic shower widths that are crucial for PFA, still exist.

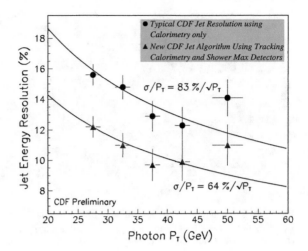

**Fig. 12.1** The effect of including information from the tracking system and the shower max detectors on the jet energy resolution measured with the CDF detector, for jets in the central rapidity (barrel) region. From: Bocci, A. et al. (2001). *Int. J. of Mod. Phys.* **A16**, suppl. 1A, 255

## 12.3 PFA at LEP, the Tevatron and the LHC

One of the conclusions of the analysis described in [6] was that the PFA approach may result in improving the jet energy resolution of "poor" calorimeter systems to become "mediocre," but that it will do little for the performance of calorimeters with "mediocre" or "good" resolution. This can be understood by considering the extreme cases: for a perfect calorimeter, there is nothing left for a tracker to improve upon, while for no calorimeter at all, the tracker would still give 30% resolution for the jets.

Practical experience so far seems to confirm this assessment. The first experiment in which PFA was elaborately applied was ALEPH [1], one of the LEP experiments. The hadron calorimeter was not considered a very important component of the LEP detectors, which had, on the other hand, excellent tracking systems. Using a specific (biased) subsample of hadronically decaying $Z^0$s at rest,[2] the authors exploited the properties of this tracking system to the fullest extent and achieved an energy resolution of 6.2 GeV, an improvement of about 25% with respect to the reported hadronic energy resolution of the stand-alone calorimeter system.

The CDF experiment at the Tevatron also used PFA techniques to improve their jet energy resolution. Figure 12.1 shows the effects of including information from the tracking system and the shower max detectors on the measured jet energy, for jets produced at central rapidities, i.e., fragments entering the calorimeter in the barrel region [2].

The CMS experiment took advantage of their all-silicon tracking system, plus a fine-grained ECAL, to improve their jet energy resolution. Figure 12.2a shows the

---

[2] Any event in which energy was deposited within 12° of the beam line, as well as any event in which more than 10% of the total energy was deposited within 30° of the beam line, was removed from the event sample used for this analysis.

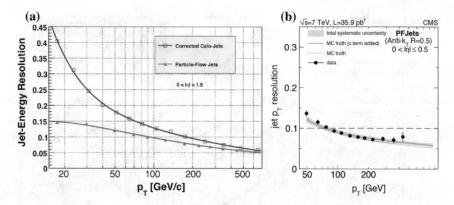

**Fig. 12.2** Simulated jet energy resolution in CMS as a function of transverse momentum, effect of PFA techniques (**a**). Measured jet energy resolution using PFA techniques as a function of energy (**b**). From: CMS Collaboration (2009). *Note* CMS-PAS-PFT-09-001

expected improvement of the jet energy resolution if PFA techniques would be used, which decreases from ~30% at 50 GeV to ~20% at 100 GeV and ~10% at 500 GeV in the energy range for which these predictions could be tested. The experimental data (Fig. 12.2b) show a somewhat smaller improvement at the lowest and highest energies measured for this purpose. However, the improvement that resulted from the use of PFA was also here certainly significant.

The ATLAS calorimeter system measures jets with much greater precision than CMS and, therefore, the replacement of the calorimeter information using the tracker data did not lead to significant benefits. However, the tracker data did help mitigating pileup effects. By subtracting the energy (momenta) of tracks that did not point to the jet vertex from the measured calorimeter energy, a better measurement of the jet energy was obtained [8].

Encouraged by the observed improvements in the jet performance, CMS has decided to replace its entire endcap calorimeter system with a dedicated PFA detector [9]. This system, which is designed to comprise about six million electronic channels, is scheduled to replace the current endcap calorimeters in the forward region around 2025. It is intended to mitigate the problems of radiation damage and event pile-up, which are expected to have rendered the current system (consisting of PbWO$_4$ crystals, backed up by a brass/plastic-scintillator hadronic section) ineffective by then. The new system will consist of $5\lambda_{int}$ deep fine-grained calorimetry, 40 sampling layers with 1–1.5 cm$^2$ silicon pads as active material, backed up by another $5\lambda_{int}$ of "conventional" calorimetry. Also the tracking system upstream of this calorimeter will be replaced, with upgrades foreseen both in granularity and in $\eta$-coverage.

## 12.4 PFA Calorimeter R&D

A large collaboration, called CALICE, has set out to test the viability of the PFA ideas. In the past 10–15 years, they have constructed a variety of calorimeters, both for the detection of em showers as well as hadronic ones. The replacement of the CMS endcap calorimeters, mentioned in the previous subsection, is based on and inspired by the work of this collaboration [5].

The calorimeters constructed by CALICE have one thing in common: a very high granularity. The calorimeter modules have a very large number of independent electronic readout channels, $\mathcal{O}(10^4)$ in most modules, up to half a million in one specific case. The active elements are either:

1. Silicon pads, typically with dimensions of $1 \times 1$ cm$^2$,
2. Small scintillator strips, read out by SiPMs,
3. Resistive Plate Chambers with small readout pads, operating in the saturated avalanche mode.
4. As an alternative, micromegas and GEMs are being tested.

These readout elements are interspersed between layers of absorber material. Typically, tungsten is used for the detection of em showers. Its Molière radius (9.3 mm) is the smallest of all practical absorber materials (only platinum is better!), so that the lateral development of the em showers is limited as much as possible. This is important for separating showers from several particles that enter the calorimeter in close proximity. For the deeper sections of the calorimeter, typically stainless steel is being used. Figure 12.3 is often shown to illustrate the advantage of using tungsten. The nuclear interaction length and especially the Molière radius, which determine the extent of the lateral shower development for hadronic and electromagnetic showers, respectively, are considerably smaller than for steel. Of course, what really matters here is the *effective* value of these parameters in the calorimeter, which also includes low-$Z$ materials such as plastic, silicon and air. A second thing to keep in mind is that the particles in reality are not colored. The difference between the colored and bitmap versions of the tungsten image illustrates that the task to assign calorimeter hits to individual jet fragments may in practice be quite daunting indeed, even in the densest possible absorber structures.

The largest calorimeter that was specifically designed for em shower detection is a tungsten/silicon device [10]. It has an active surface area of $18 \times 18$ cm$^2$ and is 20 cm deep, subdivided longitudinally into 30 layers. The first 10 layers are $0.4X_0$ thick, followed by 10 layers of $0.8X_0$ and finally another 10 layers of $1.2X_0$, for a total absorption thickness of $24X_0$. The active layers consist of a matrix of PIN diode sensors on a silicon wafer substrate. The individual diodes have an active surface area of $1 \times 1$ cm$^2$, and there are thus $18 \times 18 = 324$ calorimeter cells per layer, 9,720 in total. These are read out by means of a specially developed ASIC. Some results of measurements of em showers with this detector are shown in Figs. 6.7 and 4.1.

CALICE also built and tested a large hadron calorimeter, a sandwich structure based on 38 layers of 5 mm thick plastic scintillator, interleaved with absorber plates

$X_0 = 1.8cm, \lambda_T = 17cm$

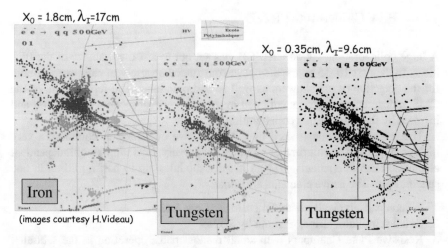

$X_0 = 0.35cm, \lambda_T = 9.6cm$

**Fig. 12.3** Simulated shower development of jet fragments in a calorimeter based on iron (left) or tungsten (center, right) as absorber material. From: Wigmans, R. (2018). *J. Progr. Part. Nucl. Phys.* **103**, 109

[11]. For this instrument, they either used 17 mm thick steel or 10 mm thick tungsten plates. This absorber material thus represents a total thickness of about $4\lambda_{int}$ in both cases. The active layers are housed in steel cassettes with 2 mm cover plates on both sides. This increased the total depth of the instrumented volume to $\sim 5.3\lambda_{int}$. The transverse dimensions of the active layers are $90 \times 90$ cm$^2$. Figure 12.4 shows a picture of one of the active layers. The layer is subdivided into tiles, small ones in the central region and larger ones in the outer regions (and also in the rear of the calorimeter module). The smallest tiles measure $3 \times 3$ cm$^2$. Each tile has a circular groove in which a wavelength shifting fiber is embedded. This fiber collects the scintillation light produced in the tile, re-emits the absorbed light at a longer wavelength and transports it to a SiPM, which converts it into an electric pulse. In total, this calorimeter contains 7,608 tiles (i.e., electronic channels). This was the first large-scale application of SiPMs in a particle detector.

Another CALICE module has a lateral cross section of $\approx 1$ m$^2$ and a similar depth as the previous one. The effective depth can be varied through the choice of the absorber material and the thickness of the absorber plates. In between each two plates an array of RPC cells with dimensions of $1 \times 1$ cm$^2$ is inserted, i.e., about 10,000 per plane [12]. In total, there are 54 independent longitudinal segments, so that the total number of active elements is about half a million. These RPCs operate in the saturated avalanche mode, and thus provide a "yes" or "no" signal when a particle develops a shower in this device. This is thus a "digital" calorimeter. An event display in this detector consists of a pattern of RPC cells that fired when the particle that created it was absorbed. These patterns may be very detailed, as illustrated by the example shown in Fig. 4.6 [13].

**Fig. 12.4** An active plastic-scintillator plane, used to detect the signals in the scintillator based CALICE hadron calorimeter. From: Adloff, C. et al. (2010). *JINST* **5**, P05004

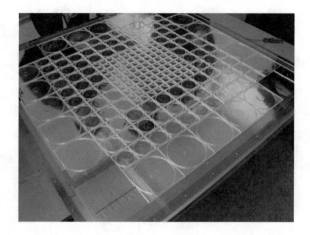

The performance of the mentioned devices as stand-alone calorimeters is not particularly impressive. For example, the em energy resolution of the W/Si em calorimeter was reported as $16.5\%/\sqrt{E} \oplus 1.1\%$ [14], which is almost twice as large as that of the ATLAS lead/liquid-argon ECAL [15]. The proposed new endcap calorimeter for CMS, which is based on this CALICE design, has an envisaged em energy resolution of $20\text{--}24\%/\sqrt{E}$ [9], an order of magnitude worse than the resolution provided by the crystals it will replace. The reasons for these poor energy resolutions are discussed in Sect. 3.1 (Fig. 6.7).

In stand-alone mode, the hadronic energy resolution of the iron/plastic-scintillator calorimeter was reported as $57.6\%/\sqrt{E} \oplus 1.6\%$, for a heavily biased event sample and after corrections that will not be applicable in a collider experiment [16]. Since this calorimeter was not deep enough to fully contain high-energy hadron showers, the event sample used to obtain this result was limited to showers that started developing in the first few layers of the calorimeter module. When this device was combined with the high-granularity W/Si ECAL, the resolution deteriorated, as illustrated in Fig. 6.14 [5].

This figure shows the signal distribution for 80 GeV pion showers before (*a*) and after (*b*) a variety of corrections were applied. The resulting energy resolution is, even after these corrections, more than three times as large as the value reported by ZEUS [17].

More worrisome than the unremarkable energy resolutions reported for these calorimeters is their non-linearity. Figure 12.5a shows the average signal of the Fe/plastic detector for positrons, as a function of energy [18]. The measured data points exhibit a significant non-linearity, namely a $\sim 10\%$ decrease of the response in the energy range from 10 to 50 GeV.

According to the authors, this is due to saturation of the SiPM signals, and they expect that this may be remedied when SiPMs with a larger dynamic range become available. Unfortunately, not enough information is supplied to verify this explanation which, if true, would also invalidate the energy resolution reported by the authors.

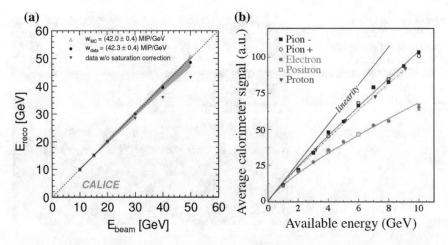

**Fig. 12.5** Non-linearity in the CALICE hadron calorimeters. Diagram **a** shows the average signal for positrons in the CALICE analog hadron calorimeter, as a function of the beam energy. Shown are the measured data points, before and after corrections for saturation in the SiPM readout were applied, as well as the Monte Carlo prediction. The shaded area represents the systematic uncertainty in the corrections. From: Adloff, C. et al. (2011). *JINST* **6**, P04003. Diagram **b** shows the average signals of the "digital" calorimeter for electrons and hadrons as a function of energy, in the energy range of 1–10 GeV. For comparison, the dependence for a linear calorimeter is given as well. Experimental data from [5, 20]

This is because, as a matter of principle, signal saturation implies that the fluctuations that determine the energy resolution are partially suppressed [19].

The signal saturation phenomenon reaches very substantial proportions in the "digital" calorimeter built by CALICE. The resulting non-linearity (Fig. 12.5b) is even so large (already for particle energies that are much smaller than expected in the experiments for which this device is intended), that it leads to apparent *over-compensation* [5, 20]. Because of the large suppression of fluctuations in the shower development process, the quoted energy resolutions are not very meaningful [19]. The large signal-nonlinearity observed for small signals is important since high-energy jets, such as the ones from the hadronic decay of intermediate vector bosons and the Higgs boson, consist of a considerable number of low-energy final-state particles, which together represent a significant fraction of the total jet energy. Quantitative information on this point is given in Fig. 7.9.

The CALICE Collaboration has apparently also realized these problems and has embarked on equipping the RPCs with a 2-bit readout system. This provides the possibility to subdivide the signals into three categories, on the basis of different threshold levels. This is called the "semi-digital" option [21]. However, the RPCs still operate in avalanche mode, and the relationship between the different thresholds (corresponding to bit settings 1/0, 0/1 and 1/1, respectively) and the deposited energy is not a priori clear.

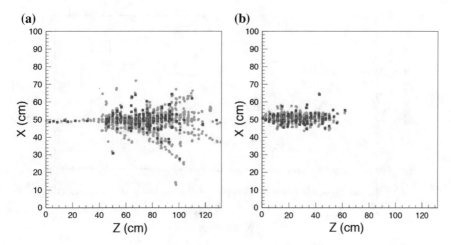

**Fig. 12.6** Event displays for a 70 GeV pion (left) and a 70 GeV electron (right) shower in the semi-digital hadron calorimeter built by CALICE. The different colors indicate the amplitude of the RPC signals (see text for details). From: Deng, Z. et al. (2016). *JINST* **11**, P04001

Some results on beam tests performed with this so-called SDHCAL are reported in [22]. The three thresholds were set at 0.11, 5 and 15 picoCoulombs in these tests. Since the average signal produced by a mip in the RPCs was 1.2 pC, the second and third thresholds corresponded to the simultaneous passage of at least 4 and 12 mips, respectively. Figure 12.6 shows event displays for a 70 GeV pion and a 70 GeV electron in which the cells exceeding the different threshold levels are indicated with different colors, red for level 3, blue for level 2, green for level 1. Not surprisingly, the average multiplicity of tracks is clearly larger for the electron shower, and the highest concentration of red cells in the pion event is found near the shower axis, where most of the $\pi^0$ production takes place.

The additional information provided in this way was used to reconstruct the shower energy. To that end the signals observed in the RPCs were given weight factors intended to compensate for the saturation effects:

$$E_{\text{reco}} = \alpha N_1 + \beta N_2 + \gamma N_3 \qquad (12.1)$$

in which the values $N_i$ represent the number of hit cells with signals above the thresholds $i$, and the weight factors $\alpha < \beta < \gamma$. It turned out that the values of the weight factors had to vary with energy in order to reconstruct the pion energy correctly. This is illustrated in Fig. 12.7. Moreover, the weight factors had to be given different values to reconstruct the energy of other types of particles (especially electrons).

Figure 12.8 shows the effects of the additional information from the RPC signals on the distribution of the reconstructed energy for event samples of pions at 20 GeV (Fig. 12.8a) and 70 GeV (Fig. 12.8b). The solid lines represent the distributions

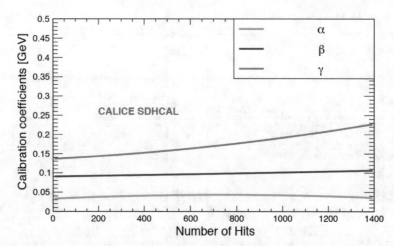

**Fig. 12.7** Evolution of the coefficients $\alpha$, $\beta$ and $\gamma$ as a function of the total number of RPC hits recorded in the CALICE digital calorimeter. From: Deng, Z. et al. (2016). *JINST* **11**, P04001

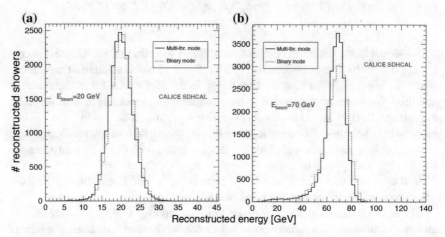

**Fig. 12.8** Distribution of the reconstructed energy with the binary mode (red dashed line) and with the three-threshold mode (solid black line), for pions of 20 GeV (**a**) and 70 GeV (**b**) in the semi-digital hadron calorimeter built by CALICE. From: Deng, Z. et al. (2016). *JINST* **11**, P04001

obtained with this information, while the dashed lines were obtained in the "binary mode", i.e., without information on the size of the RPC signals. The difference is only significant at the highest energy.

Proponents of the PFA approach argue that all these shortcomings of their calorimeters are not very important, because of the limited role played by the calorimeter in the measurement of jet properties. The most crucial feature of the calorimeters developed in this context is the very high granularity, intended to unravel the structure of the jets. Events such as the ones shown in Figs. 4.6 and 12.6, which were obtained with a cell size of $1 \times 1$ cm$^2$, are used to illustrate this point.

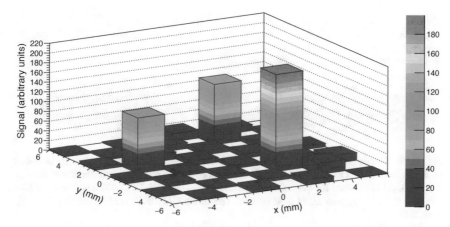

**Fig. 12.9** Event display for an event in which three particles enter a dual-readout calorimeter equipped with SiPM readout simultaneously [23]. Shown are the signals from the scintillating fibers, each of which was connected to a SiPM sensor. The Čerenkov fibers were fed through holes in the white fields and read out by a second SiPM array located directly behind the one used for the scintillation signals. The calorimeter area covered by this event display has transverse dimensions of only $1.2 \times 1.2$ cm$^2$. From: Antonello M. et al. (2018). *Nucl. Instr. and Meth.* **A899**, 52

However, this is by no means a feature that is unique for calorimeters of this type. Calorimeters that were developed to achieve excellent performance in stand-alone mode are also capable of measuring the detailed structure of events in which several particles enter the calorimeter in close proximity. Figure 12.9 shows an event display in which three particles enter a dual-readout calorimeter equipped with SiPM readout simultaneously [24]. In an area with transverse dimensions of $1.2 \times 1.2$ cm$^2$ (i.e., almost the same as the size of one RPC cell in the CALICE digital calorimeter), the three peaks are clearly separated from each other. This feature, which is a consequence of the extremely narrow shower profile (Fig. 5.6a) illustrates the benefits of a lateral detector granularity that is much finer than one would choose based on the value of the Molière radius (31 mm in this particular calorimeter).

These benefits include the possibility to distinguish between em showers caused by (the two $\gamma$s from) a $\pi^0$ and by a single electron or $\gamma$. Figure 12.10 shows simulated event displays of a 50 GeV electron shower and of a 100 GeV $\pi^0$ produced in a vertex 2 m upstream of the entrance window of such a calorimeter. The two $\gamma$s, which develop showers that are separated by only a few mm, are clearly recognizable as such.

Because of this feature and other unique advantages of the dual-readout approach (Sect. 8.3), several experiments that are being planned for future $e^+e^-$ colliders are considering a calorimeter system of this type [25].

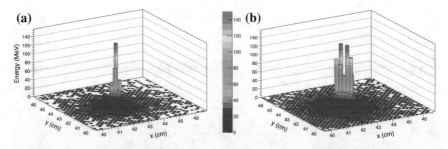

**Fig. 12.10** Simulated event displays of a 50 GeV electron shower (**a**) and the showers produced by the $\gamma$s from the decay of a 100 GeV $\pi^0$ (**b**) produced 2 m upstream of a calorimeter of the type that was used to measure the event from Fig. 12.9. From: Pezzotti, L. (2018). *Dual-readout fiber sampling calorimetry with SiPM light sensors*, Talk at the CALOR18 conference, Eugene (OR), May 2018

# References

1. Buskulic, D., et al.: Nucl. Instrum. Methods **A360**, 481 (1995)
2. Bocci, A., et al.: Int. J. Mod. Phys. **A16**, suppl. 1A, 255 (2001)
3. CMS Collaboration: Note CMS-PAS-PFT-09-001 (2009)
4. Behnke, T., et al.: (2013). arXiv:1306.6329 [physics.ins-det]
5. Sefkow, F., et al.: Rev. Mod. Phys. **88**, 015003 (2016)
6. Lobban, O., Sriharan, A., Wigmans, R.: Nucl. Instrum. Methods **A495**, 107 (2002)
7. Hadronic Shower Simulations, 6–8 Sept. 2006, Fermilab; AIP Conference Proceedings 896, eds. M. Albrow and R. Raja (2007)
8. Aaboud, M., et al.: Eur. Phys. J. **C77**, 466 (2017)
9. CMS Collaboration: Technical proposal for the phase-II upgrade of the Compact Muon Solenoid, CERN-LHCC-2015-10 (2015)
10. Repond, J., et al.: JINST **3**, P08001 (2008)
11. Adloff, C., et al.: JINST **5**, P05004 (2010)
12. Drake, G., et al.: Nucl. Instrum. Methods **A578**, 88 (2007)
13. Repond, J.: Nucl. Instrum. Methods **A732**, 466 (2013)
14. Adloff, C., et al.: Nucl. Instrum. Methods **A608**, 372 (2009)
15. Aharouche, M., et al.: Nucl. Instrum. Methods **A568**, 601 (2006)
16. Adloff, C., et al.: JINST **7**, P09017 (2012)
17. Behrens, U., et al.: Nucl. Instrum. Methods **A289**, 115 (1990)
18. Adloff, C., et al.: JINST **6**, P04003 (2011)
19. Livan, M., Wigmans, R.: Misconceptions about calorimetry. *Instruments* **1**, 3 (2017). https://doi.org/10.3390/instruments1010003
20. Repond, J., et al.: (2012). https://twiki.cern.ch/twiki/pub/CALICE/CaliceAnalysisNotes/CAN-039.pdf
21. Steen, A., et al.: (2014). arXiv:1403.8097 [physics.ins-det]
22. Deng, Z., et al.: JINST **11**, P04001 (2016)
23. Pezzotti, L.: Dual-Readout Fiber Sampling Calorimetry with SiPM Light Sensors, Talk at the CALOR18 Conference, Eugene (OR), May 2018 (2018)
24. Antonello, M., et al.: Nucl. Instrum. Methods **A899**, 52 (2018)
25. Ferrari, R.: JINST **13**, C02050 (2018)

# Chapter 13
# Outlook

## 13.1 Calorimeters and Physics Discoveries

Calorimeters have played an important role in the recent major discoveries that have shaped and improved our understanding of the fundamental structure of matter and the ways it behaves. The enrichment of the arsenal of experimental techniques brought about by calorimetry is illustrated by the fact that different features of the calorimetric performance made the crucial difference in these discoveries. As examples, we mention

- The discovery of the intermediate vector bosons [1, 2] was possible thanks to the capability of the calorimeters to recognize and select events in which large-$p_\perp$ particles (including neutrinos!) were produced on-line, amidst a background that outnumbered these events by some seven orders of magnitude.
- The observation of $\nu$-oscillation effects in Super-Kamiokande [3, 4] was possible thanks to some special features that derived from the use of Čerenkov light as the source of experimental information in this calorimeter, namely ($a$) the possibility to distinguish between low-energy electrons and muons, and ($b$) the capability to reconstruct the direction of the particles that generated the light cones.
- The experimental observation of the Higgs boson [5, 6] became possible thanks to the capability of the calorimeters to detect neutral particles ($\gamma$s) and to measure their four-vectors with very high precision.

Other important discoveries in the past half century also benefited in essential ways from calorimetry. For example, the discovery of the top quark by D0 was made possible by the capability of the calorimeter and its auxiliary systems to recognize charged leptons that were part of jets [7]. And the unraveling of the details of direct $CP$-violation in the $K^0$ system was made possible by the excellent mass resolution for $\pi^0$s in the calorimeters of the NA31 [8], KTeV [9] and KLOE [10] experiments.

In the 1950 and 1960s, a large number of new "elementary particles" were found in bubble chamber experiments, as resonances in invariant-mass plots (see Eq. 11.4). Searches for new particles were conducted by carefully measuring the four-vectors

© Springer Nature Switzerland AG 2019
M. Livan and R. Wigmans, *Calorimetry for Collider Physics, an Introduction*,
UNITEXT for Physics, https://doi.org/10.1007/978-3-030-23653-3_13

of all particles produced in the interactions and combining the four-vector of one particular particle (e.g., a $K^0$) with that of other particles (e.g., two pions of opposite charge), according to the chosen decay mode. Such searches were often successful (e.g., $K_1(1270) \rightarrow K^0\pi\pi$).

This technique is still being used successfully, for example in the spectroscopy of particles containing a charmed and/or bottom quark, in which the LHCb experiment currently excels [11]. However, as the energy at which we study the structure of matter is further increased, exclusive hadronic decay modes become very complicated (they involve many final-state particles) and each represent only a tiny fraction of the decays. For example, the decay of some charmed and bottom mesons proceeds through more than one hundred different channels with measurable branching fractions.

Increasingly, the primary process at the quark level is of great interest, rather than the precise composition of the final state. For example, the hadronic decay of a $W$ boson proceeds through the process $W \rightarrow q\bar{q}$. In order to demonstrate that hadronically decaying $W$s were produced in certain reactions, it is much more important to be able to reconstruct these two jets and determine their invariant mass than to know the characteristics of all individual particles making up the final state.

In the 1960s, $K^0$s and $\Lambda$s could be identified by the fact that the $\pi^+\pi^-$ hypothesis produced the correct mass value for the first one, while the $p\pi^-$ hypothesis reproduced exactly the 1115.68 MeV/c$^2$ on the books for the $\Lambda$. The higher the energy, the more precise the experimental measurements of the four-vectors had to be to distinguish these two particles.

History repeats itself, but now the individual particle tracks have been replaced by jets. Figure 13.1 shows the jet–jet invariant mass distribution measured by the UA2

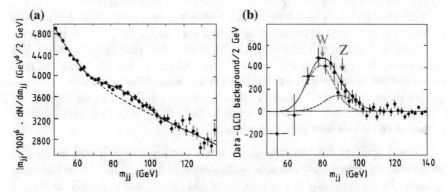

**Fig. 13.1** Two-jet invariant mass distributions from the UA2 experiment. Diagram **a** shows the measured data points, together with the results of the best fits to the QCD background alone (*dashed curve*), or including the sum of two Gaussian functions describing $W$, $Z \rightarrow q\bar{q}$ decays. Diagram **b** shows the same data after subtracting the QCD background. The data are compatible with peaks at $m_W = 80$ GeV and $m_Z = 90$ GeV. The measured width of the bump, or rather the standard deviation of the mass distribution, was 8 GeV, of which 5 GeV could be attributed to non-ideal calorimeter performance. From: Alitti, J. et al. (1991). *Z. Phys.* **C49**, 17

Collaboration [12]. The peak corresponding to the decay of the intermediate vector bosons $W$ and $Z$ can clearly be distinguished from the QCD background. This is a direct result of the measurement accuracy obtained with the calorimeter used in this experiment. An even better resolution might make it possible to observe $W$ and $Z$ decays as separate peaks in this mass spectrum. Calorimeters are ideal instruments for this type of measurement.

Thirty years ago, these intermediate vector bosons were the objects of intense experimental searches. Now, they may become the key to discovering new phenomena at the 0.1–1.0 TeV mass scale, and multi-quark spectroscopy may turn out to be an invaluable tool for this.

The quality of the hadronic calorimetry may well turn out to be very important for the study of phenomena beyond the Standard Model, assuming that there is physics beyond this Model that can be studied with the available tools. The discovery of Supersymmetry hinges on detection of the escaping LSP (Lightest Supersymmetric Particle), which manifests itself through missing (transverse) energy in the collisions. Establishing the existence of other hypothesized new objects, such as leptoquarks, requires the explicit detection of fragmenting quarks and precise measurements of their properties.

## 13.2 Lepton Versus Hadron Colliders

Matter as we know it consists of leptons and quarks. Whereas the properties of leptons such as electrons or muons can usually be measured with a very high degree of precision, the same is not true for quarks. Quarks are "locked up" inside mesons or (anti-)baryons and any attempt to isolate them creates more such particles. In high-energy scattering experiments aimed at studying their properties, quarks, diquarks or anti-quarks fragment into *jets* of hadrons.

The precision with which the properties of the fragmenting object can be measured depends on two factors: The jet-defining algorithm and the detector quality. Usually, a jet is defined as the collection of particles that fall within a cone with opening angle $R$ emerging from the interaction vertex. Typical values of $R$, when expressed in terms of an interval in $\eta$, $\phi$ space ($R = \sqrt{\Delta\eta^2 + \Delta\phi^2}$), range from 0.3–0.5. If the chosen $R$ value is large, the cone may be contaminated with particles that have nothing to do with the fragmenting object, if $R$ is small, some jet fragments may be located outside the cone. Fluctuations in the jet energy contained within the jet-defining cone form an irreducible component of the jet energy resolution.

At energies below 100 GeV, the contributions of this irreducible component are substantial and in practical experiments they are the main factor limiting the jet energy resolution. However, at higher energies, jets become more and more collimated and the effects of the jet algorithm on the energy resolution diminish correspondingly.

The relative contributions of the two factors that define the measurement precision of jets do not only depend on the energy, but also on the conditions in which these jets are produced. The effects of particles that do not belong to the fragmenting object but migrate into the jet-defining cone are extremely dependent on the type of experiment ($e^+e^-$ or $pp$ colliders, fixed target experiments) and also on factors such as the luminosity and the kinematic region under study. These *underlying event* effects are largest in the high-$\eta$ regions of high-luminosity $pp$ collider experiments, such as those at the LHC, where every interesting event is accompanied by ∼25 other events taking place in the same bunch crossing. On the other hand, in high-energy $e^+e^-$ colliders disturbances of this type are not very important.

One of the problems in designing calorimeter systems for modern experiments is the fact that the requirements for excellent energy resolution for single hadrons and jets may be orthogonal to those for high-resolution electromagnetic calorimetry. As discussed in Chap. 8, compensation is only achieved in sampling calorimeters with a small sampling fraction, which limits the energy resolution achievable for the detection of electrons and $\gamma$s. This was an important consideration that led to the development of dual-readout calorimeters, in which this limitation plays a less important role.

The effects of the jet-defining algorithm on the jet measurement precision was studied in some detail by Lobban et al. [13]. In their Monte Carlo studies, the fragmentation process is governed by a function

$$D(z) = (\alpha + 1) \frac{(1 - z)^\alpha}{z} \tag{13.1}$$

in which $D(z)$ denotes the probability that a jet fragment carries a fraction $z$ of the energy of the fragmenting object. The parameter $\alpha$ can be chosen as desired. It has been demonstrated that a function of this type gives a reasonable description of the fragmentation processes measured at LEP and at the Tevatron, for parameter values $\alpha = 3$ and 6, respectively [14].

Jet fragments are generated with energies $z E_{\text{jet}}$, with the values of $z$ chosen from a distribution representing Eq. 13.1. Each fragment is assigned a mass $m$, a charge and a transverse momentum $p_\perp$. Ten percent of the particles are assumed to be kaons and ninety percent pions. One third of the particles are electrically neutral, the rest are charged. The transverse momentum is chosen from an exponentially falling distribution with a mean value of 0.3 GeV/c. If the chosen parameters yield an unphysical result, e.g., if the chosen mass is larger than the fragment's energy $z E_{\text{jet}}$, or if the transverse momentum is larger than the total momentum $\sqrt{(z E_{\text{jet}})^2 - m^2}$, the fragment is discarded and a new one is selected. The selection of jet fragments is continued until the jet energy is exceeded. In that case, the energy of the last fragment is reduced so that the total energy of all fragments combined equals the jet energy.

The number of particles constituting the jets is quite large and increases with the jet energy. However, only relatively few particles contribute substantially to the total energy. For example, in $\alpha = 3$ jets the 10 most energetic particles carry 90% of the total jet energy. For $\alpha = 6$ jets, that takes 15 particles, on average. This is true at all energies, which is of course a direct consequence of the very concept of a fragmentation function that depends only on $z$.

The authors defined the cone parameter $R$ that formed the basis of the applied jet algorithm as

$$R = \sqrt{(\Delta\phi)^2 + (\Delta\eta)^2} \qquad (13.2)$$

where $\Delta\phi$ and $\Delta\eta$ denote the spread around the nominal direction of the fragmenting object in the azimuthal and polar angles, respectively. The fate of a jet fragment was decided on the basis of the ratio of its transverse and longitudinal momenta, $p_\perp/p_\parallel$. If

$$\arctan(p_\perp/p_\parallel) > R/2,$$

then the fragment fell outside the cone, otherwise it was considered to contribute to the measured jet characteristics (energy, momentum, composition).

Figure 13.2a shows the average fraction of the jet energy that was found to be contained in a jet-defining cone, as a function of the jet energy, for two different cone sizes, $R = 0.3$ and $0.5$, respectively. The error bars on the data points indicate the spread in the jet containment resulting from the difference in the parameter $\alpha$ used in the fragmentation function.

These data show that for jets of about 20 GeV, on average some 30% of the energy was carried by fragments that travelled outside the cone. However, as the jet energy increases, the containment rapidly improves. For energies above 100 GeV, typically less than 10% of the energy is unaccounted for when the chosen jet algorithms are applied.

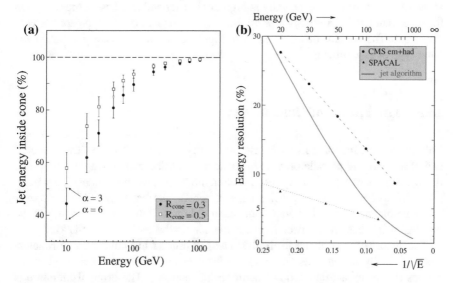

**Fig. 13.2** The average fraction of the jet energy contained in the jet-defining cone as a function of the jet energy. Results are given for cones with $R = 0.3$ and $R = 0.5$. The error bars indicate the spread in the results caused by the choice of the value of the fragmentation function parameter $\alpha$ (**a**). The hadronic energy resolution of two different calorimeter systems and the contribution of a jet-defining cone with R = 0.3 to the jet energy resolution, as a function of energy (**b**). From: Lobban, O., Sriharan, A. and Wigmans R. (2002). *Nucl. Instr. and Meth.* **A495**, 107

The energy resolution caused by *fluctuations* in the energy carried by particles traveling outside the cone is shown by the red curve in Fig. 13.2b. For jet energies of 45 GeV, as found in the decay of $Z^0$ bosons produced at the $e^+e^-$ collider LEP, the contribution to the energy resolution from jet algorithms such as those discussed here amounted to ~10%. Therefore, there was no compelling reason to install detectors measuring hadrons with a precision better than that in the LEP experiments. However, as the energy increases, the situation changes. The jets become more and more collimated and, as a result, fluctuations in the energy contained inside the jet-defining cone are reduced. For jets of 500 GeV and higher, the contribution of the jet algorithm to the jet energy resolution is of the order of 1%, smaller than the instrumental energy resolution achieved with any hadron calorimeter that has ever been tested. Therefore, a high-resolution hadron calorimeter would in practice make a crucial difference for the precision with which high-energy jets can be measured.

Figure 13.2b also shows, apart from the contribution from the "irreducible" fluctuations, the measured hadronic energy resolutions for the CMS experiment [15] and for the SPACAL calorimeter [16]. The latter calorimeter would represent a significant advantage (compared to the LHC ones) for the detection of jets with energies above 100 GeV, but at lower energies the quality of the measurements is dominated by the jet algorithm.

It should be emphasized that in these simulations, the *underlying event* effects were not included. Experience has shown that these effects dominate the energy resolution at the LHC experiments, especially at high luminosity and large pseudorapidity. On the other hand, in high-energy $e^+e^-$ colliders disturbances of this type are not very important. The conclusions derived from Fig. 13.2 are thus primarily valid for the latter type of experiments.

## 13.3   The Future of Calorimetry

Calorimeters are instruments for measuring energy. The history of physics in general, and of nuclear and particle physics in particular, is filled with examples that prove that measurement precision pays off. A better, more accurate instrument allows more precise measurements. More precise measurement results make it possible to discover new phenomena, and/or to better understand old ones. Better understanding of the physical world has always been a crucial element in the evolution of mankind.

The history of calorimetry itself illustrates this process in a nutshell. Calorimeters were originally invented as crude, cheap instruments for some specialized applications (for example, detection of neutrino interactions). The original literature is testimony to the fact that their performance was often perceived as somewhat mysterious by their users. Only after the physics on which calorimeters are based was understood in detail did it become possible to develop these detectors into the precision instruments that they are nowadays and which form the centerpiece of many modern experiments in particle physics.

We started this book with a description of the development of nuclear $\gamma$-ray detectors. The advent of germanium-based solid-state detectors with their unprecedented energy resolution caused a revolution in nuclear spectroscopy in the 1960s (cf. Fig. 1.1). We have now entered an era in which calorimetry may allow the measurement of fragmenting quarks and gluons with nuclear-spectroscopic precision. If nature is kind to us, a new world might open up as a result.

# References

1.  Arnison, G., et al.: Phys. Lett. **B122**, 103 (1983)
2.  Banner, M., et al.: Phys. Lett. **B122**, 476 (1983)
3.  Fukuda, Y., et al.: Phys. Rev. Lett. **81**, 1562 (1998)
4.  Proceedings of the 6th Topical Seminar on Neutrino and Astroparticle Physics, San Miniato (I), eds. F. Navarria and G. Pelfer, Nucl. Phys. B (Proc. Suppl.) **85**, 44 (2000)
5.  Aad, G., et al., ATLAS Collaboration: Phys. Lett. **B716**, 1 (2012)
6.  Chatrchyan, S., et al., CMS Collaboration: Phys. Lett. **B716**, 30 (2012)
7.  Abbott, B., et al.: Phys. Rev. **D60**, 012001 (1999)
8.  Barr, G.D., et al.: Phys. Lett. **B317**, 233 (1993)
9.  Alavi-Harati, A., et al.: Phys. Rev. Lett. **83**, 22 (1999)
10. De Simone, P.: J. Phys. Conf. Ser. **171**, 012051 (2009)
11. Aaij, R., et al.: Phys. Rev. Lett. **114**, 062004 (2015)
12. Alitti, J., et al.: Z. Phys. **C49**, 17 (1991)
13. Lobban, O., Sriharan, A., & Wigmans, R.: Nucl. Instrum. Methods **A495**, 107 (2002)
14. Green, D.: Dijet Spectroscopy at High Luminosity. Fermilab Report (1990). Fermilab-Conf-90/151
15. De Barbaro, P., et al.: Nucl. Instrum. Methods **A457**, 75 (2001)
16. Acosta, D., et al.: Nucl. Instrum. Methods **A308**, 481 (1991)

# Correction to: Calorimetry for Collider Physics, an Introduction

**Correction to:**
**M. Livan and R. Wigmans,**
*Calorimetry for Collider Physics, an Introduction*,
**UNITEXT for Physics,**
**https://doi.org/10.1007/978-3-030-23653-3**

In the original version of the book, Figures 5.13, 5.15 and 11.5 have been corrected in Chapters 5 and 11. The book and chapters have been updated with the changes.

The updated version of these chapters can be found at
https://doi.org/10.1007/978-3-030-23653-3_5
https://doi.org/10.1007/978-3-030-23653-3_11

**Fig. 5.13** The average
shower leakage for 100 GeV
pions and protons in the
ATLAS calorimeter system,
as a function of
pseudorapidity [4].
Experimental data from [16]

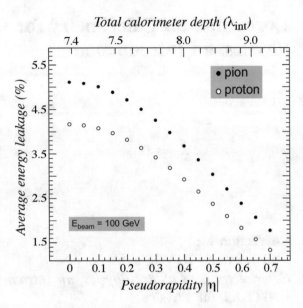

**Fig. 5.15** Signal distributions
for 75 GeV pions and
electrons in a preshower
detector used in beam tests of
CDF calorimeters [2]

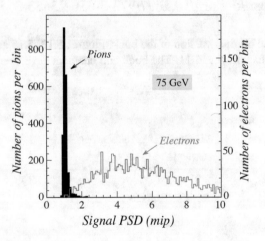

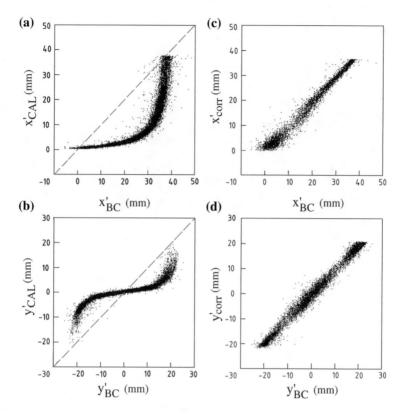

**Fig. 11.5** Scatter plots for 80 GeV electrons detected with the SPACAL calorimeter, showing the relations between the coordinates of the particle's impact point, measured with wire chambers upstream of the calorimeter (horizontal), and determined from the calorimeter data on the basis of the center-of-gravity method (vertical). The data shown in **a** and **c** concern the $x'$ coordinate, **b** and **d** refer to the $y'$ coordinate. The plots in **c** and **d** were obtained after applying the corrections from Eq. 11.2 to the calorimeter data. See text for details. From: Acosta, D. et al. (1991). *Nucl. Instr. and Meth.* **A305**, 55

# Appendix
# Some Data Relevant to Calorimetry

## A.1 Variables and Their Units

The following table summarizes the notation for variables used in this book, and the units in which these variables are usually expressed.

| Symbol | Definition | Units |
|---|---|---|
| $A$ | Number of nucleons in nucleus | dimensionless |
| $c$ | Speed of light | $2.99792458 \cdot 10^8$ m s$^{-1}$ |
| $e$ | Elementary charge | $1.6021765 \cdot 10^{-19}$ C |
| $E$ | Energy | GeV |
| $f_{em}$ | Electromagnetic fraction of hadron showers | dimensionless |
| $f_{samp}$ | Sampling fraction for a mip | dimensionless |
| $h$ | Planck's constant | $6.626075 \cdot 10^{-34}$ J.s |
| $k_B$ | Birks' constant | g cm$^{-2}$ MeV$^{-1}$ |
| $m_e$ | Electron rest mass | 511.0 keV $(=m_e c^2)$ |
| $m_p$ | Proton rest mass | 938.3 MeV $(=m_p c^2)$ |
| $n$ | Index of refraction | dimensionless |
| $N_A$ | Avogadro's number | $6.02214 \cdot 10^{23}$ mol$^{-1}$ |
| $r_e$ | Classical electron radius | $2.818 \cdot 10^{-15}$ m |
| $X_0$ | Radiation length | g cm$^{-2}$ or cm |
| $Z$ | Number of protons in nucleus | dimensionless |
| $\beta$ | Velocity as a fraction of the speed of light | dimensionless |
| $\gamma$ | Lorentz factor $(1 - \beta^2)^{-1/2}$ | dimensionless |
| $\varepsilon_c$ | Critical energy | MeV |
| $\eta$ | Pseudorapidity $(- \ln \tan(\theta/2))$ | dimensionless |
| $\theta$ | Polar angle in collider experiments | radians |
| $\theta_C$ | Čerenkov angle, $\arccos(1/\beta n)$ | radians |
| $\lambda$ | Wavelength | μm, nm, Å |
| $\lambda_{int}$ | Nuclear interaction length *for protons* | g cm$^{-2}$ or cm |
| $\lambda_{att}$ | Attenuation length for *light* | m |
| $\rho_M$ | Molière radius | g cm$^{-2}$ or cm |
| $\phi$ | Azimuthal angle in collider experiments | radians |

© Springer Nature Switzerland AG 2019
M. Livan and R. Wigmans, *Calorimetry for Collider Physics, an Introduction*,
UNITEXT for Physics, https://doi.org/10.1007/978-3-030-23653-3

## A.2　Calorimetric Scaling Parameters

Table A.1 lists parameters relevant for shower development in materials that are frequently used in sampling calorimeters: the critical energy ($\varepsilon_c$), the radiation length ($X_0$), the Molière radius ($\rho_M$), the nuclear interaction length ($\lambda_{int}$) and the specific ionization ($\langle dE/dx \rangle$) for mips. The critical energy was determined on the basis of $dE/dx$ data using Rossi's definition (see Sect. 2.2.1). The nuclear interaction length is given for *protons*, the values for pions are up to 50% larger.

**Table A.1** Properties of calorimeter materials

| Passive material | $Z$ | Density (g cm$^{-3}$) | $\varepsilon_c$ (MeV) | $X_0$ (mm) | $\rho_M$ (mm) | $\lambda_{int}$ (mm) | $(dE/dx)_{mip}$ (MeV cm$^{-1}$) |
|---|---|---|---|---|---|---|---|
| C | 6 | 2.27 | 83 | 188 | 48 | 381 | 3.95 |
| Al | 13 | 2.70 | 43 | 89 | 44 | 390 | 4.36 |
| Fe | 26 | 7.87 | 22 | 17.6 | 16.9 | 168 | 11.4 |
| Cu | 29 | 8.96 | 20 | 14.3 | 15.2 | 151 | 12.6 |
| Sn | 50 | 7.31 | 12 | 12.1 | 21.6 | 223 | 9.24 |
| W | 74 | 19.3 | 8.0 | 3.5 | 9.3 | 96 | 22.1 |
| Pb | 82 | 11.3 | 7.4 | 5.6 | 16.0 | 170 | 12.7 |
| $^{238}$U | 92 | 18.95 | 6.8 | 3.2 | 10.0 | 105 | 20.5 |
| Concrete | – | 2.5 | 55 | 107 | 41 | 400 | 4.28 |
| Glass | – | 2.23 | 51 | 127 | 53 | 438 | 3.78 |
| Marble | – | 2.93 | 56 | 96 | 36 | 362 | 4.77 |
| Active material | $Z$ | Density (g cm$^{-3}$) | $\varepsilon_c$ (MeV) | $X_0$ (mm) | $\rho_M$ (mm) | $\lambda_{int}$ (mm) | $(dE/dx)_{mip}$ (MeV cm$^{-1}$) |
| Si | 14 | 2.33 | 41 | 93.6 | 48 | 455 | 3.88 |
| Ar (liquid) | 18 | 1.40 | 37 | 140 | 80 | 837 | 2.13 |
| Kr (liquid) | 36 | 2.41 | 18 | 47 | 55 | 607 | 3.23 |
| Xe (liquid) | 54 | 2.95 | 12 | 24.0 | 42 | 572 | 3.71 |
| Polystyrene | – | 1.032 | 94 | 424 | 96 | 795 | 2.00 |
| Plexiglas | – | 1.18 | 86 | 344 | 85 | 708 | 2.28 |
| Quartz | – | 2.32 | 51 | 117 | 49 | 428 | 3.94 |
| Lead-glass | – | 4.06 | 15 | 25.1 | 35 | 330 | 5.45 |
| Air 20°, 1 atm | – | 0.0012 | 87 | 304 m | 74 m | 747 m | 0.0022 |
| Water | – | 1.00 | 83 | 361 | 92 | 849 | 1.99 |

## A.3 Scintillating Materials Used in Particle Physics Experiments

The tables in this subsection concern materials that are used for particle detection in physics experiments. Relevant properties of some scintillating crystals and plastic scintillators used for this purpose are listed in Tables A.2 and A.3, respectively.

**Table A.2** Properties of scintillating crystals that are used in particle physics experiments

| | NaI(Tl) | CsI(Tl) | CsI | BaF$_2$ | CeF$_3$ | BGO | PbWO$_4$ |
|---|---|---|---|---|---|---|---|
| Density (g cm$^{-3}$) | 3.67 | 4.51 | 4.51 | 4.89 | 6.16 | 7.13 | 8.30 |
| Radiation length (cm) | 2.59 | 1.85 | 1.85 | 2.06 | 1.68 | 1.12 | 0.89 |
| Molière radius (cm) | 4.8 | 3.5 | 3.5 | 3.4 | 2.6 | 2.3 | 2.0 |
| Interaction length (cm) | 41.4 | 37.0 | 37.0 | 29.9 | 26.2 | 21.8 | 18.0 |
| $\langle dE/dx\rangle_{\mathrm{mip}}$ (MeV cm$^{-1}$) | 4.79 | 5.61 | 5.61 | 6.37 | 8.0 | 8.92 | 9.4 |
| Refractive index (at $\lambda_{\mathrm{peak}}$) | 1.85 | 1.79 | 1.95 | 1.50 | 1.62 | 2.15 | 2.2 |
| Hygroscopicity | Yes | Slight | Slight | No | No | No | No |
| *Emission spectrum, $\lambda_{\mathrm{peak}}$* | | | | | | | |
| Slow component (nm) | 410 | 560 | 420 | 300 | 340 | 480 | 510 |
| Fast component (nm) | | | 310 | 220 | 300 | | 510 |
| *Relative light yield* | | | | | | | |
| Slow component | 100 | 45 | 5.6 | 21 | 6.6 | 9 | 0.3 |
| Fast component | | | 2.3 | 2.7 | 2.0 | | 0.4 |
| *Decay time* (ns) | | | | | | | |
| Slow component | 230 | 1300 | 35 | 630 | 30 | 300 | 50 |
| Fast component | | | 6 | 0.9 | 9 | | 10 |

**Table A.3** Properties of some plastic scintillators and wavelength shifters applied in particle physics experiments. Listed are the wavelengths at which the absorption (for WLS) and emission spectra peak, the characteristic decay time and, when available, the absorption length (WLS)

| | Absorption (nm) | $\lambda_{abs}$ ($\mu$m) | Emission (nm) | Decay time (ns) |
|---|---|---|---|---|
| *Scintillators* | | | | |
| SCSF-38 (Kuraray) | | | 428 | 2.3 |
| SCSF-81 (Kuraray) | | | 437 | 2.4 |
| SCSF-3HF (Kuraray) | | | 530 | 7 |
| BCF-10 (Bicron) | | | 432 | 2.7 |
| BCF-20 (Bicron) | | | 492 | 2.7 |
| BCF-60 (Bicron) | | | 530 | 7 |
| *Wavelength shifters* | | | | |
| Y-7 (Kuraray) | 440 | 100 | 490 | 10 |
| Y-11 (Kuraray) | 435 | 80 | 476 | 10 |
| BCF-91A (Bicron) | 420 | | 494 | 12 |
| BCF-92 (Bicron) | 410 | | 492 | 2.7 |

## A.4 Noble Liquids Used in Calorimeters

Table A.4 lists relevant properties of noble liquids used in particle physics experiments.

**Table A.4** Properties of noble liquids applied in particle physics experiments

| Property ↓ Liquid → | LAr | LKr | LXe |
|---|---|---|---|
| Z | 18 | 36 | 54 |
| Boiling point (K) | 87.3 | 119.8 | 165.0 |
| Density in liquid phase (g cm$^{-3}$) | 1.40 | 2.41 | 2.95 |
| Radiation length (cm) | 14.0 | 4.7 | 2.40 |
| Molière radius (cm) | 8.0 | 5.5 | 4.2 |
| Nuclear interaction length for protons (cm) | 84 | 61 | 57 |
| *Ionization properties, measured with $^{207}$Bi* | | | |
| Energy needed per electron–ion pair (eV) | 24 | 17 | 15 |
| *Scintillation properties, for 1 MeV $e^-$ ($\alpha s$)* | | | |
| *Emission spectrum, $\lambda_{peak}$ (nm)* | 128 | 147 | 174 |
| *Decay time* (ns) | | | |
| Fast component | 5.0–6.3 | 2.0 | 2.2 |
| Slow component | 860–1090 | 80–91 | 27–34 |
| *Relative light yield* | | | |
| Fast component | 8% (57%) | 1% | 5% (31%) |
| Slow component | 92% (43%) | 99% | 95% (69%) |

Printed in the United States
By Bookmasters